The Art of Forgetting

Ivan Izquierdo

The Art of Forgetting

 Springer

Ivan Izquierdo
Memory Center
Pontifical Catholic University of Rio
    Grande do Sul, and National Institute
    of Translation Neuroscience
National Research Council
Porto Alegre, Brazil

ISBN 978-3-319-06715-5   ISBN 978-3-319-06716-2   (eBook)
DOI 10.1007/978-3-319-06716-2

Library of Congress Control Number: 2015931258

Springer Cham Heidelberg New York Dordrecht London

Printed on acid-free paper

Springer International Publishing AG Switzerland is part of Springer Science+Business Media
(www.springer.com)

Foreword

We all know a lot about forgetting because we do a lot of it. And, as Izquierdo discusses and emphasizes throughout this little book, it is mostly good that we forget. After all, most of our experiences are, well, forgettable. We do not need to recall for our lifetimes, if at all, our uncountable daily experiences. But, we do need to maintain memories of important experiences as well as the skills that help us to survive. Memory is, for all animals, critical for survival. If memory is so critical, however, why do we so readily forget most of our daily experiences?

In an engaging style that mixes historical ideas with literary observations and experimental findings, Izquierdo discusses what is known and what needs to be known about the battle between remembering and forgetting. The studies of Ebbinghaus, Pavlov and other pioneers of the late nineteenth century revealed much about the conditions of practice that influence learning and subsequent memory. Research mostly of the past half century provided a rich and detailed understanding of different forms of learning and memory.

More recent studies of the past several decades, including those from Izquierdo's laboratory, have revealed much about how the activity of neurons and brain systems enables us to learn and remember. Although less is known about why we forget, Izquierdo discusses much of what is known as well as why it is important to understand forgetting. As he notes, both incorrect remembering and failures to forget can form the bases of distressing clinical problems.

Before proceeding to turn to the first page of this book, the reader is advised to make sure that she/he has no other pressing obligations, as interesting ideas are found on every page. Izquierdo has written a highly engaging, thoughtful and informative book. Much of it is likely to be unforgettable.

Irvine, CA, USA

James McGaugh
Research Professor of Neurobiology and Behavior
Center for the Neurobiology of Learning and Memory
University of California Irvine

Preface

This is the English version of the second edition of a book published in Portuguese in 2011. Many changes were introduced here relative to that former version. I thank Drs. Jociane de Carvalho Myskiw and Cristiane Furini for their many valuable suggestions as well as the many readers of the former versions for their comments and support. The artwork of the book is by Dr. Myskiw and Ms. Maria Eduarda Izquierdo.

Porto Alegre, Brazil   Ivan Izquierdo

Contents

Introduction

"In fact, Funes not only remembered every leaf from every tree of every group of trees, but also every time he had perceived it or imagined it."

Jorge Luis Borges,
"Funes the Memorious,"
First published in *Ficciones* (*Fictions*), 1944 [1].

Memory is defined as the storage and retrieval of information by the brain, and forgetting is defined as the loss of memory, which is called amnesia.

Jorge Luis Borges described in his famous short story the fictitious case of a young Uruguayan peasant who after a fall from a horse acquired an "absolute" or "total" memory; he could remember absolutely everything that happened to him or that he did or saw, including each and every detail, like the peculiar shape of the clouds on a given afternoon. In those years memory was considered to be a rather mysterious process of which much was said and who nobody dared to study biologically.

Perhaps the person who knows more about memory today is my friend James (Jim) McGaugh, who was actually one of the first to study the biology of memory when the field began in the 1950s: he discovered that there was a definable process of making memories in the brain, called consolidation, that will be commented in several sections of this book. In 1971 McGaugh published with Harry Harlow and Richard Thompson the textbook *Psychology* [2] in which he said that "the most salient aspect of memory is forgetting." Interestingly, Jim recently turned to the analysis of people with superior autobiographical memory (like Funes) [3, 4]. Jim McGaugh and I frequently talk about memory and forgetting, including our own. We have dedicated most of our lives to the study of memory, which gives us many remembrances to ponder about.

I agree with his statement about forgetting; it is obvious to me that most of what happened to any of us in the last 30 min, or yesterday afternoon, or in the past

© Springer International Publishing Switzerland 2015

I. Izquierdo, *The Art of Forgetting*, DOI 10.1007/978-3-319-06716-2_1

10 years, or through all our life is irretrievably forgotten. From a lively conversation
we had with a colleague half an hour ago, we remember only the general aspects but
few if any complete phrases. From yesterday's events, aside from those that were
truly very important, we remember very little. Indeed, we usually need no more than
a few minutes to remember all we can about yesterday morning or any day last
week. We sometimes cannot remember the words of a song we used to sing some
years ago. Gone are, as if they never existed, the names, faces, persons, knowledge,
events, numbers, and details of our whole life. Eventually fortuitous circumstances,
a cue, a word, or some look-alike person can bring back memories we thought were
lost; but really there is no trace to be found of most of them, aside from dubious
fragments modified by time and by nostalgia, nothing more than fleeting and imperfect reminiscences.

Many interviews with hundreds of people, among them several considered to
have a good memory, attest to the fact that there is more forgetting than memories
in our minds, as McGaugh pointed out in 1971. An exception is the rare individuals
with highly superior autobiographical memory recently described by McGaugh and
his coworkers mentioned above [3], but even in them some of the memories may be
false [4]. The Uruguayan writer Mario Benedetti (1920–2009) published a book
called *El olvido está lleno de recuerdos* (*Forgetting is full of memories*) [5]. The
Spanish film director Luis Buñuel wrote perhaps the most moving description of
human memories and of the subtleties and properties of their spotty remembrance
in the very old, based on his own vivid forgetting and on the terminal forgetting of
his senile mother: *Mi último suspiro* (*My last breath*) [6].

Jim McGaugh and I have good memories, and we sometimes talk about our
childhood or about shared experiences or about memory research data we both
know. Childhood is the period of life in which we all learn some of the very fundamental things, including walking; speaking; reading; writing; swimming; how to
ride a bike; how to recognize the aspect, smells and sounds of people, animals and
things, the tactile sensation of flowers, wood and animals, and the song of birds; and
many of the facts, preferences, and abilities that are crucial to our of life, including
love and how to choose whom we love. We learned the difficult arts of deciding
between doing and not doing, responding and not responding, and fighting or flee-
ing and of distinguishing the main from the accessory, the useful from the useless,
and when to be hopeful and when not to be, as it says at the entrance of Dante's hell.
Many episodes in our infancy left key memories that will unconsciously be important for us for the rest of our life: the goodnight kiss of our mother; her tenderness
when she picked us up after a fall; the warmth and smell of her breast when we were
very little; the peculiar vibration of our father's chest when he spoke; the smile of
our parents, brothers, sisters, and friends; the feeling of the fur of our first dog; the
sounds and smells of our home and neighborhood; and the sense of the past and
therefore of continuity provided by our grandparents, in short, memories that established the foundations of our affective, sentimental, and cognitive world. But usually we cannot remember the moment in which each of those memories was first
established, whether it was in winter or summer; if it was when we were 2, 4, or 12
years old; or if they were acquired in single or in repeated episodes. We all easily
mix up the various events of our lives, even those of greater significance, including

the exact content and circumstances of each one of them. As years go by, the face of our first grade teacher mingles with that of the second or third grade teacher or with that of an aunt or a friend.

Early in life, some say that in intrauterine life, we learn usually from our mother the accent and the peculiar cadence of the language that we correctly call our mother language. If I speak with the particular lilt I have in any language, it is perhaps because I learned from my mother, who was born in Croatia when it belonged to the Austro-Hungarian Empire and was educated in Argentina, a mix that eventually produced the singsong that became typical of me, whatever that may be.

Once well out of the uterus and in control of our own little but eager hands and feet, we learn first the meaning or purported meaning of sounds or syllables (a, ma, ba, pa), then that of isolated words, and then their integration into phrases, sentences, and eventually paragraphs. But we do not remember exactly the first phrase we ever spoke or the moment when that happened. We do not remember the crucial evolutive moment in which we crossed the line between isolated words and their association in phrases. That moment would distinguish us forever among other species and as individuals within our own species.

If anybody asks us to tell everything we remember about our childhood, we can probably do so in a couple of hours or less. We all lost valuable memories: the faces of our first grade schoolmates and episodes that were once significant and now we cannot connect with any longer. Sometimes images or words from our infancy reappear in dreams, but we cannot recognize them. In adulthood, we often forget details of the face of our first girlfriend or our grandparents, which were (and are) so important to us. Often we feel an almost physical pain when we fail to remember those faces. Our life involves the loss of neurons and synapses from our first year on, and important memories may lie in those lost neurons and synapses.

Borges cites Gustav Spiller [7] as stating that remembering all the memories of an intelligent 60- or 70-year-old person would probably take no more than a day and a half or two. If we ask the best physician in the world to tell us all he/she knows about medicine, he/she will probably do it in one morning. In order to go beyond that, he/she would need his/her library, his/her computer, and perhaps the help of some of his/her colleagues. Borges himself, once when asked about his first memory, mentioned a vague sunset; but in all probability it must have been some untraceable and also vague pain in the belly, or the warmth and smell of the breast that fed him, or feeling cold for the first time. Newborns pay little attention to sunsets.

Clearly, each day many of our memories disappear forever. But in spite of this, we all function reasonably well, communicate with each other using language, understand what is going on in our surroundings, and live in more or less organized communities in which many complex systems operate and function under our control. We all know a lot of information that enable us to survive. Some of us are taller, poorer, thinner, or wiser, but each one of us is "somebody," an individual. Each of us is who we are because each of us has his/her own memories. As the great Italian philosopher, Norberto Bobbio (1909–2004) said, we are strictly what we remember [8]. In this short book I intend to show that we are also what we learn or choose to forget.

What we forgot without a trace is as alien to us as if we had never known it; it is no longer in our brains and therefore not a part of us. What we do not know does not belong to us. We all make plans for the future, short or long, based only on what we remember from our own past, even if we have only fragments of it. We simply cannot make plans based on what we forgot or on what we never knew.

Thus, forgetting may indeed be the most salient aspect of memory, but we all keep enough memories or memory fragments to behave actively and more or less satisfactorily as persons. We remember who we are, where we live, where we work, and who are our relatives and friends. The loss of those memories is incapacitating, as in the advanced stages of the diseases that affect memory, such as the dementias (from the Latin *de* (privative particle) and *mens, mentis* (mind)). That is why dementias are so desolate and why normal lives, however spare they may be at a very old age, are not. Many neurons and their synapses die in very old age, and that leaves us with fewer memories when we get to be 90 or 100. But as is known our life expectancy grew in the past decades, thanks to the advances of sanitation, hygiene, medicine, and other health sciences, prominently including pharmacology. So the advanced age at which memories begin to fade has gone up along with our life expectancy. One finds many old people to be mentally active and productive at 80, 90, or more years of age today, particularly those dedicated to intellectual pursuits, as will be seen in a later section of this book; this is because reading is the best known way to practice memory.

When the brain of an old person does not fall victim to a neurodegenerative disease, tumors, or trauma, it remains functional and able to store more and more memories every day, and so his or her usefulness to society does not decrease. In fact, it may actually increase with age, because a very old person can bring us first-hand knowledge of things of many years ago which would otherwise remain unknown.

With the increase of life expectancy that has taken place in the last half-century because of the advances in sanitation and medicine, the decline in brain function begins every year a little later. Just take a look at the increasing median age of some great performing artists, scientists, writers, and particularly Nobel Prize and Oscar winners over the past couple of decades. Of course there are parts of the world in which public health is still appalling and people die in their infancy, but the percentage of people who live better and longer now than 40 or 50 years ago has nevertheless increased enormously over most of the globe. When I came to live in Brazil in 1973, the life expectancy of Brazilians was 51 years; now (in 2014) it is over 75 (79 for women).

There appears to be something selective and purposeful in our increased forgetting as years go by. We all forget many things but not those that at bottom really matter. We normally do not forget who we are, where we live, and what we do for a living,except when our brains get demented. If our brain stays healthy, we simply make more memories every day and just go on piling them up, no matter the age. "With age, my brain has become like a circus full of people where more people come in every day," a very old man once told me. The brain of most humans practices a kind of art when it permits (or promotes) the forgetting of so many memories, while keeping most of the main ones intact in order to keep the circus going. It behaves as most cities do when centuries go by. Some buildings become ruins, but enough of them persist so they can be repaired and other new ones may be built on top of them so as to replace them. But the streets, many of the major buildings, and water reservoirs remain in place in many of the world's great cities over the ages, and so does our memory each year at a more advanced age. Our forgetting is not similar to the result of indiscriminate bombing but much more like the effect of the passage of time on cities, and our minds retain a lot of their distinctive character much like Athens, Rome, Paris, or even the much newer Rio does over the years. Dementia is much more like Hiroshima or Nagasaki in the first days after the bomb.

Time turns not only our minds but also our bodies into ruins. The art consists of keeping the noblest parts (brain, mind, liver, heart, arteries, lungs, and kidneys) alive and active till the end. All that is left for us to do is to rebuild on the ruins, like the Romans or the Greeks before us did. By doing precisely that, Queen Elizabeth I, Deng Xiaoping, Giuseppe Verdi, and Jorge Luis Borges were able to go on and even change their styles after the once remote age of 80 and performed with excellence at their respective jobs: to govern a complex and enormous empire, to build a great new country, to compose great music, and to write great literature, building anew over their old and perhaps partly dilapidated memories. They may be old and partly dilapidated, but they gain some new items every day while they lose others. It is easier for the old to keep on working with their memories nowadays than it was a few years ago, given the vast amount of peripheral systems that have become available: big libraries, all sorts of recording and computer devices including cell phones, Internet-based search programs, etc. It was much more difficult to gather information and build upon it in the old days and still is so in primitive societies, where word of mouth is the only means of information storage and communication.

Finally, a comment about the insertion of memory studies in science as a whole today. In the twentieth century, the branch of science dedicated to the study of the morphology, properties, and functions of nervous systems, including our own, has come to be known as Neuroscience. The fathers of Neuroscience are considered by most to be the Spanish histologist, Santiago Ramón y Cajal (1852–1934); the English physiologist, Charles Scott Sherrington (1857–1952); and the Russian physiologist, Ivan Petrovich Pavlov (1849–1936). Cajal (as he is usually called in English) studied carefully the entire neuroanatomy of the brain of humans and vertebrates; what we know of neuroanatomy today is largely the fruit of his research. Sherrington discovered the main laws of the entire physiological organization of central nervous systems, which is based on reflex activities whereby the stimulation of one neuron affects the activity of others. Pavlov was the first to approach the study of learning, memory, and forgetting from a biological point of view. He used the concept of the separateness of neurons, which was established by Cajal, and the concept of reflexes, which was developed earlier by his Russian precursor, Ivan Mikhailovich Sechenov (1829–1905), and then prominently by Sherrington. Prior to the work of these pioneers and their numerous disciples, behavior was explained by hypothetical constructs with no real link to definite brain processes, except for the also pioneering work of the American philosopher and psychologist William James (1842–1910), seen by most as the founder of modern psychology, of which he raised most of the fundamental questions. Several of these were of biological nature, and some were answered in the century that followed him by biological methods.

Neuroscience englobes neuroanatomy, neurophysiology, neurochemistry, neuropsychopharmacology, and neuropsychology and collaborates with other fields among which chemistry and computer science have been very prominent in the past 20 or 30 years. The most currently used types of psychotherapy and all pharmacotherapy have their roots in Neuroscience, including the various forms of cognitive therapy. This permitted the amenability to treatment of a large variety of mental ailments previously outside the range of any form of therapy, including schizophrenia, autism, dyslexia, depression, anxiety, posttraumatic stress disorder, phobias, and addictions. A biological basis for all these disturbances is now known and can be acted upon.

The scientific study of memory is now clearly within the scope and activities of Neuroscience, unlike a hundred years ago, when it was believed to belong to psychology, then a somewhat ill-defined branch of philosophy. The progress of Neuroscience in the past 50 years has been phenomenal. It grew from scratch into one of the fastest-growing sciences that attracts more young scientists. Many brain mechanisms regulating behavior are now known or at least well understood, including those of learning, memory, and forgetting. A series of new scientific pursuits have emanated from Neuroscience in the past decade or two, including neurolinguistics, neuroengineering, neuroeconomics, and neuroeducation; they apply the laws and principles of Neuroscience to other domains of human activity.

References

1. Borges JL. Ficciones. Buenos Aires: Emecé; 1944.
2. Harlow H, McGaugh JL, Thompson RF. Psychology. San Francisco: Albion; 1971.
3. LePort AK, Mattfeld AT, Dickinson-Anson H, Fallon JH, Stark CE, Kruggel F, Cahill L, McGaugh JL. Behavioral and neuroanatomical investigation of highly superior autobiographical memory (HSAM). Neurobiol Learn Mem. 2012;98:78–92.
4. Patihis L, Frenda SJ, LePort AK, Petersen N, Nichols RM, Stark CE, McGaugh JL, Loftus EF. False memories in highly superior autobiographical memory individuals. Proc Natl Acad Sci U S A. 2013;110:20947–52.
5. Benedetti M. El olvido está lleno de memoria. Montevideo: Cal y Canto; 1995.
6. Buñuel L. Mi último suspiro. Barcelona: Plaza y Janés; 1994.
7. Spiller G. The mind of man; a text-book of psychology. London: S. Sonnenschein; 1902.
8. Izquierdo I. Qué es la Memoria? (trans., What is memory?) Buenos Aires: Fondo de Cultura Económica; 1992. 92 pp.

The Art of Forgetting

2

2.1 The Formation and Retrieval of Memories

From the point of view of their contents, there are several types of memory. Those that everybody calls memories are called declarative. They comprise the memory of meanings, understandings, and concept-based knowledge, called semantic memory, and the memory of episodes, called episodic or autobiographical memories. Typical examples of semantic memories are a language, chemistry, medicine, psychology, or a symphonic score. Typical examples of episodic memories are a language class, a movie, or something that happened to us or we heard about. Episodic memories are also called *autobiographical* because we remember them only inasmuch as they refer to ourselves; we cannot remember a movie seen by somebody else. The memories of sensory or motor skills are called procedural memories or habits (how to use a keyboard, how to ride a bicycle). Learning how to play a flute is procedural; learning what to play in the flute is semantic. French classes are episodic; knowledge of the French language is semantic.

A special form of memory that will be treated in the next section is working memory, now viewed as a mostly online system of keeping data on the essential properties and the context of what we are learning and/or retrieving at any time and where we are doing that. When we ask somebody for a telephone number, we keep the number "in mind" for a few seconds so as to dial it, and then we lose it altogether.

Most, if not all, declarative memories are acquired and formed in the region of the temporal lobe called hippocampus (Fig. 2.1), a phylogenetically old cortical structure, with the help of the much newer neighboring cortical region through which the hippocampus connects with the rest of the brain, called entorhinal cortex, and that of the also neighboring, rounded nuclear complex located in front of it, called the amygdala (in Latin, little almond). All declarative memories are then stored in various cortical areas (see below).

Procedural memories are made in the cerebellum and in the so-called basal ganglia, of which the most important for this is the caudate nucleus. The memories of addictions, which are believed to be largely procedural even though during craving

© Springer International Publishing Switzerland 2015
I. Izquierdo, *The Art of Forgetting*, DOI 10.1007/978-3-319-06716-2_2

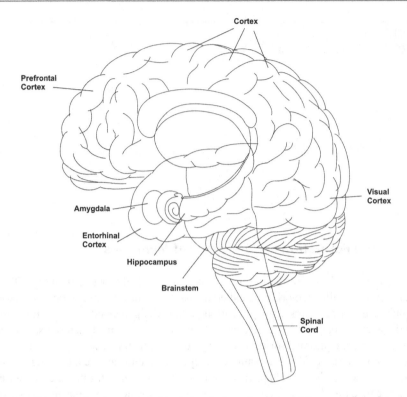

Fig. 2.1 Schematic drawing of the left brain hemisphere showing, as if by transparency, the location of some of the main structures involved in memory formation (hippocampus, amygdala, and entorhinal cortex) located within the temporal lobe, as well as other nervous structures (prefrontal cortex, rest of the cortex, brain stem, cerebellum, and spinal cord). The hippocampus, amygdala, and entorhinal cortex are involved in declarative memories and the basal ganglia (not shown here) and cerebellum, in procedural memories or habits

they elicit delusions and sometimes deliriums (which comprise declarative components), are made and stored in the nucleus accumbens, anatomically associated with the basal ganglia. The amygdala itself, which also plays a role in some aspects of addiction, is ontogenetically and anatomically related to the basal ganglia too [1].

From the point of view of their duration, all the various types of memory are classified as short term (6 h or less) and long term (over 6–24 h). Long-term memories that persist for many weeks, years, or even decades, like those of our childhood, are often called remote. Most long-term memories last only 1 or 2 days; unless repeated and fortified by reconsolidation (see below), they simply fade away.

The making of memories involves first of all a process of acquisition, which is what most people call learning, and consists of the translation of what we perceive in experiences of all sorts into brain language, which consists of biochemical and electrical signals intertwined (see below). The experiences through which we learn can have an external origin (what we perceive with our sensory systems), an internal origin (what we perceive through our interoceptive systems, such as a pain in the stomach or a full bladder), or a "mental" origin (another memory, the association of things present with others in our past, predictions of things that have not happened yet). Of course the number of experiences that can cause learning may be finite but is certainly unlimited, and so is the number of memories we can possibly make. Once experiences are translated into brain language, the resulting information is consolidated into what are called memory traces or files. The consolidated information is stored in synaptic networks in different parts of the brain and is kept in brain language. The process of consolidation involves first a process that takes up to about 6 h and occurs mostly in the hippocampus and is called cellular consolidation. Starting simultaneously in cortical regions, first in the entorhinal cortex and then in other areas, another much longer-lasting process develops called systems consolidation, which extends beyond cellular consolidation for many days, months, or even years [1, 2]. The discovery of the very long-term systems consolidation process began by observations on the amnesic effect of repeated electroconvulsive shocks on memories acquired many months or years ago in [2] and was corroborated by numerous observations on the spontaneous incorporation of false details into long-standing memories (see below). Systems consolidation is believed to consist of complex neuronal rearrangements involving the growth of new neuronal processes and synapses, as well as the atrophy of others. The physiological connection between cellular and systems consolidation is still being worked out, but it is known that it does involve the hippocampus [2].

The retrieval of memories requires both the structures in which they were originally formed and various regions of the cortex, including the posterior parietal cortex, which has been known for years to be involved in knowing where we are located in space at any given time [1]. The mechanisms whereby these particular brain areas are able to reassemble the learned information stored and codified in brain language back into that other language we use to communicate with what we call reality are not known.

As said, available evidence suggests that most memories are forgotten during our lifetime. This leads to several questions.

The first is "why do we forget?"

We forget probably in part because the mechanisms that make and retrieve mechanisms may be saturable. They cannot be put to work simultaneously all the time for all possible memories, those that already exist and those that are acquired every minute. This requires that preexisting unused memories be put away to make place for new ones.

We do not know exactly the mechanisms used by the brain to store the *main* elements of each memory. We do not yet know even what those main elements are. But we know quite a lot about the biochemical steps that regulate such mechanisms, and that permits us to figure out important aspects of memories [1]. Experimental evidence suggests that storage occurs through structural and functional modifications in some of the synapses involved in the making of each memory [1, 2], as was postulated by one of the founders of modern Neuroscience, the

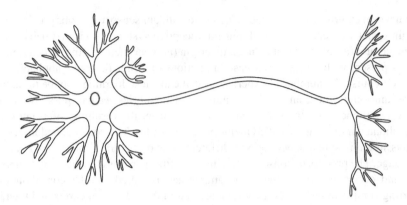

Fig. 2.2 Schematic drawing of a prototypical brain cortical neuron. Many of these are called "pyramidal cells" because their cell body may look a bit like a pyramid. The long prolongation to the *left* is called the *axon* and carries spikes from its initial segment where they are generated to its termination(s) at synapses on other nerve cells or on other cells (muscular, glandular). The *axons* are less ramified than the prolongations that emanate from the cell body to the *left*, which that are called *dendrites*; these can receive the end feet of 1,000–10,000 axon terminals each, which enable the nerve cell to receive a much more numerous amount of messages brought by spikes from other nerve cells than those it is capable of emitting to other neurons

great Spaniard Santiago Ramón y Cajal, in 1893 [3]. Apparently the brain-derived neurotrophic factor (BDNF), a peptide produced in many brain areas, plays a key role in this. BDNF is produced in the hippocampus, where many memories are made, and in the neocortex, where many of them are stored, in different phases of memory formation [1]. Other neurotrophic factors include one called nerve growth factor, which was the first of them to be described, and another one produced in glial cells (the non-nervous cells that surround and support neurons in nervous tissue, assisting them in nutritional activities or in the disposal of metabolites [1]). There is evidence that memories change the shape and number of synapses in many brain systems [4, 5].

There are over 80 billion neurons in the human brain. Some of those in the cortex and hippocampus receive between 1,000 and 10,000 axonal terminals (Fig. 2.2). Thus, the possibilities of functional intercommunications between brain cells are enormous, and each of these synapses may participate in many different memories. Thus, the informational capacity of the brain is immense, particularly that of the hippocampus and those regions of the rest of the cortex that makes memories and is involved in their maintenance.

The second question is "what do we forget for?"

The response comprises several different aspects and levels of analysis, and I will attempt to provide at least partial answers in the following pages. Borges in "Funes the Memorious" said that we need to forget in order to think, for which we need to make generalizations [6].

2.2 Forms of Forgetting: First Steps of the Art (Extinction, Repression, and Discrimination)

By "forgetting" we usually mean not being able to remember memories. There are basically four ways of not remembering memories: habituation, extinction, differentiation (also called discrimination), and repression. The first three are forms of learning.

All of them consist of making memories less accessible without erasing them. They look in many respects like forgetting, and even some neuroscientists view them as forms of forgetting. Indeed, for most practical purposes, they constitute what most people call forgetting: the memories simply "are not there when called." They do not involve or require erasure of the memory files but consist, rather, on the inhibition of the responses to stimuli normally used to retrieve them.

Some people outside the field of Neuroscience believe that memories are kept in the brain in little packages that can be singled out and detached individually, like books on a shelf. This is not the case, however, and one cannot extricate one single memory from all the others it connects with, in time or over time. The memory of a perfume comes together with that of our mother or our first love, or both, or that of those two persons plus other perfumes. The memory of Ohm's law comes to me together with the face of my physics teacher in high school and that of my mentor in my second postdoc year in UCLA 50 years ago, John Green.

So aside from time itself, which can no doubt delete many memories (most of yesterday afternoon's memories or last year's, as mentioned above), there is very little that can be done to do away with memories except training ourselves not to remember them.

The "art of forgetting" concentrates on the use of these four forms of not remembering (*habituation*, *extinction*, *differentiation*, and *repression*) and, as will be seen, also on memory falsification. It is, therefore, not an art of real forgetting, but one of learning how to use to our advantage the processes that the brain has to inhibit remembering. We realize that we have a memory only when we retrieve it; otherwise it lies beyond our reach. The art of forgetting is, alas, as imperfect as any other art, and it does not consist in promoting or preventing the real erasure of memories, which the brain does on its own without any participation of our will most of the times to our benefit.

The simplest process for inhibiting retrieval is habituation. It was first described by Ivan Petrovich Pavlov in the first few years of the twentieth century [7] and is best explained through examples. The first time we enter a room, we usually look around until we become more familiar with it. The first time we hear a sound or are exposed to the flash of a light or see somebody entering the room, we turn our head in the direction of the stimulus. These natural responses are what Pavlov called the "what is it?" reflex and most scientists call the "orienting reaction." They are performed by all animals in response to novel stimuli, and in dogs, cats, or other animals with a better olfactory or auditory system than ours, they are performed together with a turn of the ears and nose towards the stimulus.

With repetition the orienting response gradually diminishes its intensity and eventually fades away. This usually gradual inhibition of the response is called *habituation* and is the simplest form of learning. Once we habituate to a stimulus, the response to it may stay inhibited for years. The first time we hear a buzzer, we startle and turn our head to where the buzzer comes from. The nineteenth time we hear it, we just ignore it. Habituation can be cancelled by an interpolated stimulus or by a change of the intensity of the original stimulus: if the sound is played louder or if its tone becomes different, or the light is made more brilliant, or the room we enter is now full of baggage. In spite of its simplicity, habituation requires a long chain of biochemical events in the hippocampus just like any other kind of learning [1].

Once habituation is established, the "what is it?" reflex may reoccur without previous notice; this is called spontaneous recovery. The simple change of context may also recover the orienting response; this is called "renewal." For example, if we habituate to a whistle in a room and then we hear it again in another room, the orienting response usually reappears.

The "what is it?" reflex has an obvious defensive character and helps us to survive. If we do not respond to a novel stimulus or situation with at least looking at it or listening to it, we could miss detecting a danger of some sort. Habituation helps us go around the world more comfortably. This is the case of people who have to work in noisy places like an airport or in places with many lights like theaters or with many voices like a public market. It is also obvious in the case of children to learn to go to school every day even if the first day came as a surprise entry into an unknown world, full of new noises, faces, places, and other stimuli. Patient H. M., whom we refer to in another section and had bilateral surgical lesions of the hippocampus and surrounding tissue, had great problems in habituating to new stimuli in a lasting way. Habituation learning is indeed regulated by the hippocampus, as will be commented further on.

In order to understand the other three major forms of the art of forgetting, we will proceed with a brief description of a form of learning also described by Pavlov, conditioning.

2.3   Conditioned Reflexes

The major contribution of Pavlov to Neuroscience was his discovery of conditioned reflexes, known all over the world as conditioning [7].

If a neutral stimulus (a context, a sound, a light, a smell) is paired repeatedly with what Pavlov called a "biologically significant" stimulus that always evokes a response (food, water, pain, feeling sick), the response to the former will become conditioned by its association with the latter and will change. For example, a tone paired with the presentation of food will cause a dog to respond to the tone with salivation. At that point the tone will have become a conditioned stimulus, the food can be called an unconditioned stimulus, and the acquired salivation response to the tone will be a *conditioned reflex*. The procedure can be used with an enormous variety of stimuli; pairing of the tone or a light with a footshock will elicit a leg flexion response to the tone or the light as a defensive conditioned reflex, and the procedure will be called by many as fear conditioning, because in humans it is associated with a sensation of fear. This type of conditioning, being the first described (in the last years of the nineteenth century), is called classical or Pavlovian. It is actually the type of learning most widely used, especially for studies on memory and forgetting because of its simplicity and its rapid acquisition (see [8–15]).

If delivery of the unconditioned stimulus is made to depend on the conditioned response, the *conditioning* will be called *instrumental*, because the response will be an instrument to bring about the unconditioned stimulus if it is food or to avoid it if it is a footshock or a bad taste. Instrumental conditioning was reported at about the same time by the Polish physiologist Jerzy Konorski in Pavlov's laboratory in Saint Petersburg and by the American psychologist Burrhus F. Skinner in Minnesota. Konorski initially called it "Type 2 conditioning." Skinner called it operant because in his setups the animals performed the instrumental response by actually operating levers or other gadgets that caused the delivery of food or the avoidance of a footshock [8]. Eastern Europeans are more prone to classifications; Americans are more adept to gadgets.

The unconditioned stimulus is called by many the reinforcement, because it is used to reinforce conditioned behavior. If once conditioning is established, the reinforcement is omitted; the animals will inhibit the expression of the conditioned response. This is the form of learning called *extinction*. The extinction of classical conditioning is faster than that of instrumental conditioning.

Discrimination or differentiation is the inhibition of the response to stimuli that are qualitatively similar to one that causes a biologically significant response. Again it is best explained by an example. Animals trained in a conditioning routine can learn, for example, that a 10 kHz tone will be followed by meat and learn to salivate to that tone in anticipation of the meat. If the animals are exposed to other tones, say, an 11 or a 15 kHz tone, it will also salivate to these other tones the first few times that they are presented. This is called generalization. But if none of these other tones is followed by meat, the animals will soon cease to do so and salivate preferentially or only in response to the 10 kHz tone, which is always followed by meat. *Generalization* is a very common phenomenon and is indeed the same that happens when a small baby starts to call everybody around "dad." Soon the baby learns to restrict the use of this word only to designate his or her real father, to his mother's relief. Inhibition of learned responses to the inappropriate stimuli (salivation to the other tones or using the word "dad" to designate any man) is called differentiation or discrimination and is learned.

Pavlov was, along with Ramón y Cajal and with the British Physiologist Charles Sherrington (see above), one of the founding fathers of Neuroscience. His discovery of habituation, the conditioned reflexes, and extinction signaled the beginnings of the experimental analysis of behavior and inaugurated the attempts to look for its neural basis.

Perhaps needless to say, habituation and *differentiation* are two pillars of the art of forgetting. Without habituation, life would be an incomprehensible collection of sensory stimuli, and we would tend to respond to all of them all the time. Without differentiation, we would not be able to distinguish a meaningful thing from others. They are essential to life as we understand it. The better we learn how to habituate and how to discriminate a significant stimulus (or a significant piece of information) from all others, the better off we will be.

There are two other ways of not remembering memories that consist of real losses of information, real forgetting, and are not part of the art of forgetting. One is by preventing memory consolidation, by erasing memories as soon as they are acquired. A head trauma or anesthetics or other drugs that depress the brain given right after acquisition can do this [9]. A knocked-out boxer may not remember, when he wakes up, in what round he was knocked out. An anesthetized patient does not remember the surgery he was submitted to. An alcoholic may not remember things he did when drunk.

The other way to do away with memories is by erasing them long after they have been consolidated, like in Alzheimer's disease or in our forgetting what we did yesterday morning. Real forgetting is not an art; it is a pity. No doubt necessary, as Borges said, but a pity nevertheless: "forgetting is necessary in order to think, to make generalizations" [6]. Without the forgetting of everyday trivial memories, our minds would be ebullient with useless information at all times. But to erase memories is philosophically a pity, because the brain dedicates an effort to make and keep each memory, in which energy is consumed and many pathways in the brain are activated.

Conditioning, classical or instrumental, is routinely used in all human activities, from the promotion of soldiers to corporals, to the first kiss, to learning how to play the flute.

It is interesting to note that in today's society, which has become so automatized, there are some who, by ignorance, dislike and even deny the existence of conditioned reflexes. They think that reflexes are automatic and that we humans, being thinking beings, are not just automatic machines. We certainly are not that, and reflexes are not as automatic as those people think. Even the "behavior" of a coin-operated vending machine can be regulated by a variety of procedures, among them blocking their input or their output with a screwdriver. Conditioned reflexes can be modulated by a variety of procedures too, such as getting tired or angry or more or less willing to perform them or having a drink or feeling good or bad [1, 8, 9]; see Sect. 2.9 further on Memory consolidation and Memory and emotion. Yes, we are humans after all, and neither us nor any other living being is automatic, but we all have conditioned reflexes. We know how to look for water or food, we make sex, and we sing the national anthem using conditioned reflexes. People who think conditioned reflexes do not exist should go back to high school biology classes and learn—for which they will have to use to a large extent conditioned reflexes, of course.

2.4   The Rapid Forgetting of Working Memory Is Intrinsic to Its Nature

As mentioned in a preceding section, there are several types of memory according to their function and duration.

First of all there is working memory, which is partly "online" and lasts just a few seconds or minutes beyond the cessation of an experience. We use it to understand the reality around us and to effectively form or retrieve other forms of memory: those that last for several minutes or hours (short-term memory, of which there is probably more than one type [10]) and those that last days, years, or decades (long-term memory). Those that last longest are often called remote memories. For the oldest of us, the memories of our childhood are remote. Memories of a day or a week ago are simply long term.

A good example of working memory is the fourth word of the preceding phrase. When I was writing it or when you were reading it, its fourth word was "kept in mind" just long enough to make sense of the phrase; in a few seconds it was gone forever, and if it hadn't, it would have confused any further reading or writing. Another example of working memory is when we ask for a phone number, remember it just long enough to use it, and then forget it altogether. If it were not forgotten for good right away, it would confuse the acquisition or retrieval of all the other numbers we were to use that day, including other phone numbers and the password to our bank account.

Older people tend to recall remote memories better than newer ones. Borges said that this is because those were the memories of "the days of happiness," when we were young and could dance all night, knew the words to all the current songs, were so often in love, played ball quite well, and thought we could change the world. The more recent years are, for the old, "the days of losses": we lose people we love, as well as our physical prowess and sometimes our health year by year (heart disease, hypertension, diabetes, and arthritis are more common in the old). It is certainly nicer to remember the happy days of our youth than the recent days of different ailments and the loss of dear friends or relatives. When recent memory is lost beyond this "benign" level of selectivity, it may signal, of course, the onset of brain disease and may be incapacitating: one may lose the ability to go to places, to hold conversations, or to do simple tasks.

Some psychological studies suggest that the older we get, the more prone we will be to remember the good things and/or to refurbish the others as being better than they really were.

Another reason why many old people remember bygone days better than the more recent ones is not as benign as Borges thought. It is because in old age the functioning of working and short-term memory systems and the retrieval of recent memories often is not as good as it was at younger ages. In most cases this loss is not sufficient to qualify it as a form of pathology. In others (perhaps in 20 % or so of people over the age of 75) it may be part of a more encompassing memory disorder called minimal cognitive impairment (MCI) (see further on).

Working memory does not form lasting files; it disappears in seconds or, at the most, minutes, so the information it contains does not interfere with preceding or subsequent information. It relies on the electrical activity of the anterolateral prefrontal cortex and to a lesser extent other prefrontal areas. When the activity of those neurons ceases, a number of seconds after the stimulus is over, working memory also ceases. Some prefrontal neurons start firing at the end of a given stimulus; these are also called "off-responses" and they signal the end of a sensory or sensory-motor experience. They are also seen in the sensory cortical areas and signal the cessation of stimuli [11, 12].

As will be seen in a later section, neurons are activated by chemical signals and transform these into electrical signals generated in their axons, which are conducted to their terminations at synapses, and are recordable as bursts of rapid potential changes called "firing."

When the mechanism of working memory is set into action by each experience, the information processed by the prefrontal cortex is communicated to other regions of the brain and exchanges data with them by comparing the respective rates of firing. These regions include those that analyze sensory information and those that store information in longer-lasting memory systems. This exchange of data is rapid and permits the brain to realize, among other things, if the incoming information is new and important and needs to be saved or is already known and can be ignored or deleted. This permits the analysis of information acquired simultaneously or at very short intervals. So we can detect melodies or use each particular word of a given phrase within a larger context, such as the understanding of a longer text, like the whole paragraph or the entire page. So we can also distinguish the man leaning against the wall across the street from other people that pass by and from the wall itself. We discriminate between the car going down the street and the trees that remain fixed while it passes, and we can estimate its speed. We compare that with preexisting information and realize what is new and what we know already.

The nature (trains of electrical signals lasting seconds produced by prefrontal cortical cells) and function of working memory make it short [12]. This correlation between nature and function may be viewed as a form of art, which we share with all living beings with a brain and a working memory mechanism. This mechanism is thought to be better in other animals than it is in humans. In humans it is quite good: it permits us to think, understand, and do things in the midst of the flood of information our senses are submitted to most of the time.

What happens when working memory fails? The power to discriminate between consecutive or simultaneous pieces of information becomes lost. Of course simultaneous information is in many if not most cases perceived by us by rapid scanning, so we really deal with it as if it consisted of closely consecutive information. When working memory fails, we cannot distinguish very well the man leaning against the wall from those that walk by or from the wall itself. Teachers with a failure of working memory will tend to perceive the class as a mass with many faces, and the question raised by a given student will mix with that asked by another and with some noise in the corridor. Smells, shapes, and sounds get mixed. Reality becomes incomprehensible to varied extents, if not outright hallucinatory. A glimpse of this may be perceived when we get exhausted, when we lose many hours of sleep, or when we are bombarded by an excess of information and stress.

Failures of working memory leading to false or deformed perceptions of reality (hallucinations) are seen in schizophrenia. Lesions of the prefrontal cortex and hippocampus have been described in this disease, and many psychiatrists view it as secondary to those anatomic defects [13].

The long-term memory of the hallucinatory states systematizes them in deliriums, many of which accompany schizophrenic patients through life. Careful psychotherapy, along with drugs that selectively reduce hallucinations and deliriums, may slowly help them to discriminate between the delirious world of their own, which they cannot share with anyone, and reality, which they share with the rest of the world. A very good, albeit somewhat schematic vision of this appears in Ron Howard's 2001 movie *A Brilliant Mind*, starring Russell Crowe.

2.5   Brain Areas and Systems Involved in the Different Types of Memory: Some Basic Notions of Neuronal Function

As commented in the preceding section, working memory results mainly from the operation of neurons of the anterolateral prefrontal cortex and other parts of this cortex. The hippocampus plus the neighboring cortex that directly projects to it or receives fibers from it, the rest of the cortex, and some other brain regions are in charge of other types of memory.

The hippocampus is a phylogenetically old region of the temporal lobe cortex that in humans is rolled into it a bit to end up a bit like the sausage is in a hot dog (Fig. 2.1). The hippocampus has several functions: the main one is to make and retrieve memories and to lead the rest of the cortex to participate in this, starting by the region closest to it, the entorhinal cortex, which is less old phylogenetically speaking. The entorhinal cortex is located right below the hippocampus, and it is linked to it and to the rest of the cortex by bundles of axons. It is also linked to the amygdaloid nucleus complex, which is composed of several subnuclei, looks a bit like an almond (in Latin, *amygdala* means little almond), and lies in front of the hippocampus like a ball that has just been kicked in front of the foot that kicked it. So the entorhinal cortex has two-way connections with the rest of the cortex, the amygdala, and the hippocampus. Several subnuclei of the amygdala register, and react to, emotions, and the basolateral nuclear complex of the amygdala is considered to be the part of the brain that "translates" emotions, particularly emotional arousal, and sends that information to the hippocampus and to the rest of the brain [1, 9, 14].

The hippocampus and its connections are the main regions involved in the formation and in the retrieval of memories. The amygdala adds emotional arousal to memory processing like one adds salt and pepper to food during cooking. Memory formation is believed since McGaugh (1966) [15] to consist basically of a process called consolidation, by which the signals registered by the brain during the process of learning or acquisition (terms used as synonymous by cognitive neuroscientists) are put together and transformed into specific chemically triggered changes at the dendrites or other postsynaptic sites of neurons that generate local graded small potential changes called postsynaptic potentials, a few millivolts (mV) high and several milliseconds long [1] (see below).

Figure 2.2 shows a neuron with the general shape of the main cells of the cerebral cortex, including the hippocampus, called pyramidal cells because of the shape of their body. Nerve cells emit two types of prolongations: one, longer and usually single, which ramifies at a distance from the cell body and carries the output of the neuron to other cells and is called the axon and several others, much shorter and

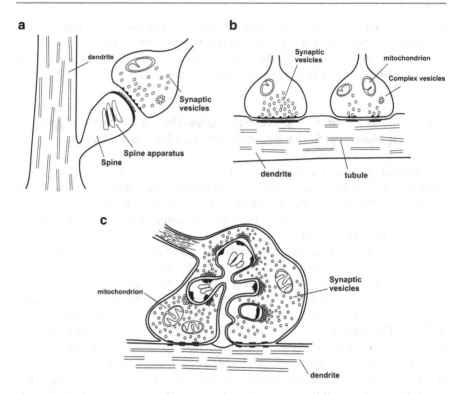

Fig. 2.3 Synapses of different shapes between the presynaptic terminals of axons and dendrites of the postsynaptic cells. The terminals are enlargements of the axons and contain mitochondria and synaptic vesicles (the vesicles that carry the neurotransmitters, which vary according to the synapse). In some cases the latter appear as more or less globular complexes. At the terminal membrane, there is, in some of them (**a**, **c**), a thickening of the membrane; on the *opposite side* of the synaptic cleft, there is another thickening (**b**, **c**) visible as a darkening of the postsynaptic membrane, where the postsynaptic receptors are. Clearly some synapses are built so as to ensure transmission "no matter what," as in (**c**), or perhaps in (**b**), where two axon feet make consecutive contact with a dendritic branch. Others, like that in (**a**), allow for a more punctiform contact (one axonal presynaptic terminal to one dendritic "spine"; spines are short branches of dendritic ramifications that usually make synapse with just one axonal terminal)

highly ramified, which receive the axon terminals of other nerve cells (the input of the neuron) and are called dendrites. The axon terminals end at a very short distance from the dendrites (100–150 Å, an angström being a hundred-millionth of a centimeter). This area of approximation between both is called the synapse, and the space between the axon terminal and the dendrite is called the synaptic cleft (Fig. 2.3). Neurons communicate with each other at synapses by small-molecule substances released from the axonal terminal called neurotransmitters which cross the cleft to interact with specific proteins on the dendrite called receptors (see below). In the synaptic terminal of the axons, these molecules are stored in vesicles called synaptic vesicles. The spikes, when arriving at the axonal terminal, cause the

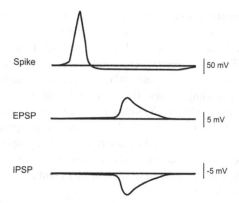

Fig. 2.4 Schematic drawing from a typical *spike or action potential* (*upper*), of a typical excitatory postsynaptic potential (EPSP) (*middle*), and of a typical inhibitory postsynaptic potential (IPSP) (*lower*). In all cases the potentials are drawn against a *horizontal straight line* that represents the neuron's resting potential, usually 60–90 mV negative with respect to the cell's extracellular fluid (this depends on the structure being analyzed). Note that the spike and the EPSP decrease this potential, thereby *depolarizing* the neuron, whereas the IPSP increases this potential, thereby *hyperpolarizing* it. The spike is at least 10 times higher than the postsynaptic potentials and a bit narrower (1–2 ms for the spike and 3–6 ms for the PSPs). Some IPSPs may actually last much longer (500 ms). Also note the long hyperpolarizing "afterpotential" that follows the spike, caused by the reposition of sodium, potassium, and chloride ions to their resting levels following the large changes caused by the spike

entry of calcium into it, which mobilizes the vesicles to the membrane on the cleft where they release their content (Fig. 2.3). Once they reach the membrane of the dendrite of the next cell, the neurotransmitters interact with receptors to cause ionic or chemical changes. The ionic changes include the so-called postsynaptic potentials, which are caused by small (<5 mV) currents carried by sodium (Na^+) or chloride (Cl^-) ions entering or potassium (K^+) ions leaving the cells at the synaptic site lasting a few milliseconds (ms). These potentials are propagated all over the membrane of the dendrites and the cell body as if it were a cable and reach the initial segment of the axon of this cell, which can generate action potentials or spikes. A *spike* is a high (around 100 mV), brief (1–2 ms) potential change that can be conducted with no decrement to the endings of the axon whose initial segment generated it (Fig. 2.4). This form of conduction is called "all-or-none" and allows for the preservation of the full 100 mV spike amplitude from the initial segment to the ending(s) of the axons at presynaptic terminals impinging on other neurons (or muscle fibers or gland cells).

Postsynaptic potentials that consist of a net influx of positive ions (Na^+) will tend to carry the polarity of that segment to the threshold for the generation of spikes and thereby cause that cell to be excited and to communicate with others and are called excitatory postsynaptic potentials (EPSPs). The threshold for triggering spikes is at a positive level relative to the resting potential of the neurons. In synapses in which the postsynaptic potentials involve a net outflux of K^+ or a net influx of Cl^-, it will tend to stabilize the initial segment of the axon at a negative level, away from the threshold, and be inhibitory to that cell; these are called inhibitory postsynaptic potentials (IPSPs) (Fig. 2.4). Neurons can communicate with other neurons or with muscles or gland cells only by sending spikes to presynaptic terminals (excitation), and when they stop sending spikes, the intercommunication is inhibited. EPSPs (Na^+ influx) increase, and IPSPs (Cl^- influx or K^+ outflux) decrease, the possibility that the initial segment of the axon (Fig. 2.2) will reach the spike threshold or stay away from it. The initial segment of the axon has electrical properties different from those of the cell body or the dendrites. Like the rest of the axon, it has no receptors to neurotransmitters, cannot generate postsynaptic potentials, but responds instead to the local entry of positive ions with

Each hippocampal pyramidal cell receives an input from 1,000 to 10,000 axons from other neurons. Some axons make synapse with cell bodies rather than dendrites, particularly those that make inhibitory synapses.

As said, the interaction of the neurotransmitters with the dendrites causes either chemical or electrical changes in the latter. The chemical changes include interactions of the neurotransmitter-receptor complex with other proteins called G proteins because they couple to the cyclic guanosine phosphate nucleotide (cGMP), and initiate a series of metabolic changes in the cell, leading to their eventual excitation or inhibition. Receptors with a G protein are called metabotropic receptors.

Neurons can communicate with other neurons or with muscles or gland cells only by sending spikes to presynaptic terminals (excitation), and when they stop sending spikes, the intercommunication is inhibited.

The series of events from synaptic potentials to synaptic transmission, if repeated, is coded into neurons by processes involving protein synthesis by ribosomes, which relies on gene translation, as well as by extra-ribosomal protein synthesis (see below) [1]. Protein synthesis is believed to underlie long-term memories and is conspicuously absent from working and short-term memory processes. The proteins synthesized in ribosomes or in the extra-ribosomal system after learning experiences are a major element of the memory-making process or consolidation. They are components of synapses or the enzymes that make them or put them in place, and they are transported to the synapses that had been recently activated much as bricks are transported to a construction site. The chain of events from synaptic transmission to protein synthesis involves, at several stages, signaling by enzymes called protein kinases, which catalyze their own phosphorylation and that of many proteins, from receptors to neurotransmitters, to nuclear transcription factors mediating the activation of genes, and to components of the extra-ribosomal protein synthesis apparatus located in dendrites [1, 8]. Phosphorylation means the incorporation of phosphate molecules to the proteins, which usually increases their function.

Working memory is believed to rely just on electrical activity [11, 12] in areas of the prefrontal cortex. Short-term memory is believed to rely mainly or only on protein kinase-mediated protein phosphorylation events in the hippocampus and its neighboring entorhinal cortex [1, 2]. Protein synthesis inhibitors administered into specific brain nuclei do not affect working or short-term memory, but block long-term memory. This indicates that long-term memory formation is, and the other, shorter forms of memory are not, dependent on the neuronal machinery for making new proteins.

Two important characteristics of synaptic potentials are their long-term potentiation (LTP) and long-term depression (LTD). These two processes occur following the sudden and brief (1–10 s) rapid (4–100/s) stimulation of axons that make synapses and are seen in many places of the brain with glutamatergic transmission, notably in the hippocampus where they were observed for the first time. They are believed to be at the basis of memory consolidation and extinction. Together they are usually known as forms of neuronal plasticity. LTP and LTD usually last for several hours, but occasionally they have been shown to last for many days or even weeks; there is one reported case in the literature in which LTP lasted at least 1 year. It was thought at the beginning, soon after their discovery, that LTP or LTD could actually represent the memories themselves and that these were stored in the usually hippocampal cells showing LTP or LTD. Now, these are more commonly seen as correlates or signatures of processes underlying memory consolidation. The places in which LTPs or LTDs last longer (the hippocampal formation, Fig. 2.1) are not involved in the preservation or retrieval beyond a few hours after acquisition [16, 17].

Neurons, like other cells, produce new proteins through two different sets of processes. One depends on the activation of genes in the cell nucleus to produce messenger RNAs that flow to organelles in the cytoplasm called ribosomes, where the messengers get together with local ribosomal RNA in order to stimulate a series of enzymes that produce proteins. The DNA in the genes is transcribed to specific messenger RNAs (mRNAs), and these are then translated into specific proteins at the ribosomes. There is another protein synthesis system first described in tumoral cells which was then found in neuronal dendrites. It relies on preexisting mRNAs and on a system in which the key element is called mTOR (an acronym for "mammalian target of rapamycin," which is the name of an antibiotic that binds to it and causes its inhibition) (see [18]). This system conglomerates around the mTOR molecule a series of enzymes which include protein kinases and elongation factors for the manufacture of new proteins. The process is fostered by the translation of preexisting messenger RNAs in the absence of any new DNA transcription. The dendritic location of the mTOR system makes it a good candidate for the expression of proteins in response to locally generated synaptic activity. Both the ribosomal and the mTOR systems are involved in the production of proteins relevant to different forms of learning, among which are fear conditioning, fear extinction, and object recognition [18–23].

2.6   Executive Functions

As our capacity to process working memory is very large but not infinite, in our cell phone- and tablet-laden modern society, we often feel suffocated or half-drowned by an excess of information. The large amount of parallel and consecutive bits of information we receive all the time nowadays requires an almost continuous use of working memory. In the 1920s Cajal himself felt this and expressed it beautifully in a book called *El mundo visto a los 80 años* (*The world seen at the age of 80*). He complained about the flood of information as perceived in those years: the sounds from the crystal radios of the neighbors which we did not ask to hear, bands marching in the street, people speaking loudly all over, too many movies in the theaters all over town, etc. He felt that all this was an invasion of the privacy of city dwellers. He foresaw a future when the large amounts of concurrent information produced by the world around us would make it difficult to concentrate, let alone rest. Cajal could not predict over 90 years ago the extraordinary variety of noises, sounds, news, pseudo-news, cries, shouts, visual and acoustic pollution devices, and other stimuli of all sorts that impinge upon us from every side nowadays. There were of course no computers, cell phones, tablets, television, or high-power sound amplifiers in his day.

Back in the 1920s, Cajal, like most people after a certain age, probably felt nostalgic for "the good old quiet days of happier times" when the world was not invaded by so many noises as it is today. This yearning for the "good old quiet days of yore" is in reality a wish for the return of a time which never existed for Cajal or for any of us 90 years later, a desire to go back in history to many generations past. Until a few centuries ago, landowners, who were a tiny minority of the population, might have been able to lock themselves up in their manors or castles and quietly breathe the silence around them. That silence was limited to an environment sealed by strong walls, often made of stone. For the vast majority of the knaves, slaves, or beggars outside the castle, life was not so quiet or silent; in many countries the slashing of the slave driver's whip frequently did away with whatever peace of mind its victims had been able to attain. When we yearn for a distant past, we usually picture ourselves as members of the upper classes, and not as the members of the lower classes, who in many places were slaves and in all cases were the overwhelming majority. In all probability most of us are descendants of the poor, simply because they were far more numerous and bore more children than the landlords. In most countries there were at least nine poor for every rich person. Fear and hunger were more important distracting factors for those people than the excess of noise. If we go further back in history to the age of the cavemen, it is likely that we would have met with excess information in the form of stressful and outright dangerous stimuli, such as wild animals, natural disasters, or unfriendly neighbors who wanted to snatch our food or our women from our caves.

In order to survive, we need to somehow rationalize the activity of our central executive of information which, by virtue of its properties, is precisely the prefrontal cortex [12]. Through its connections with the rest of the brain via the entorhinal cortex, that region can detect right away whether a given ongoing experience belongs to those that we have filed as dangerous memories or to those of a milder nature. The excess of information can overload the system. In order to gain some time for carrying out that process of rationalization, one useful procedure is to learn to say "no!" when confronted with excessive demands on our working memory. If we cannot attend to the demands of a given person or situation because we are taking care of other persons or situations, we must first quickly appraise the priority of the new stimulus as opposed to that of the others. Ideally, we should learn to say "no!" politely in order to stop, think, and assign priorities, but there are many urgent cases in which we do not even have the time to say "no." If we are in the middle of an important meeting in a room and we are interrupted by a band of thieves or by fire, our catalogue of priorities is immediately overruled: our only priority is to get out. In airplanes the crews recommend us to adjust our own oxygen mask first before helping whoever is sitting at our side to adjust his or her mask. If we lose our consciousness, we will not be in a position to help anybody else.

Older people generally develop a better and more comprehensive catalogue of priorities over life than younger people; they have had more time and many more experiences to assist them in this endeavor. Unfortunately the very old may not be in good physical conditions to implement that catalogue (sensory deficits, slower reflexes). Learning to what kind of information we should give priority over others is a difficult art. It requires to be instantly prepared to set aside temporarily most of the demands that are being placed on us, along with all previous priority lists, when we are confronted with something of more pressing importance. The example of the oxygen masks mentioned above or that of the band of thieves are quite illustrative as metaphors.

2.7 More on the Connections Between Nerve Cells

In previous sections we have mentioned the general function of synapses. In many parts of the nervous system, axons make synapse with the dendrites or cell body of the neurons. In the case of the projections of the nervous system to the rest of the body, the axons make synapses with muscles or with gland cells and make them contract or relax (in the case of the muscles) or secrete hormones (in the case of the glands).

As said, the chemically mediated postsynaptic potentials generated at dendrites result from the interaction of different neurotransmitter substances with postsynaptic proteins called receptors. The most common excitatory neurotransmitter is glutamic acid, and the most common inhibitory neurotransmitter is γ-amino-butyric acid (known for its acronym, GABA). Other major brain neurotransmitters are acetylcholine, norepinephrine, dopamine, serotonin, and histamine. Acetylcholine mediates neurotransmission at the peripheral parasympathetic system and neuromuscular junctions. In the former it produces inhibition (bradycardia, lowering of blood pressure, inhibition of intestinal contractions), and in skeletal muscles it causes excitation (the contraction of those muscles). Norepinephrine mediates sympathetic nerve transmission; dopamine mediates transmission in many areas of (i.e., those involved in sexual stimulation, pleasant behavior, the satiation of needs, or even the effect of drugs that may bring about real or imagined well-being). Serotonin regulates mood, and low serotonin function correlates with the symptoms of depressive illness. All these neurotransmitters plus histamine (which in the periphery causes allergic vascular reactions, urticarial rashes, and itching) modulate memory processing in the hippocampus, amygdala, and prefrontal cortex through actions mediated by a variety of different receptors.

Neurotransmission by GABA is called GABAergic. Neurotransmission by glutamate is called glutamatergic. Synapses in which the neurotransmitter is acetylcholine, norepinephrine (in most English-speaking countries called adrenalin or adrenaline), dopamine, serotonin, and histamine are called cholinergic, noradrenergic, dopaminergic, serotoninergic, and histaminergic, respectively. In all cases there are several different receptors for each transmitter; for example, there are two different receptors for GABA, called $GABA_A$ and $GABA_B$. There are α, β, and γ noradrenergic receptors; some of them are located in presynaptic axon terminals, regulate the release of norepinephrine by those terminals, and are called α_2 and β_2. There are four different receptors to histamine, called H1, H2, H3, and H4; the five varieties of dopaminergic receptors are called D1, D2, D3, D4, and D5. The interaction of each transmitter with each receptor yields different effects; the action of norepinephrine at α receptors causes an elevation of blood pressure and tachycardia; that upon β receptors causes a lowering of blood pressure and bradycardia.

The term "neurotransmitters" is usually reserved for substances that act on neighboring receptors. There are other substances released at axon terminals that act at a distance on receptors distributed over larger areas, binding to receptors in many neurons, gland cells, or vascular smooth muscles; these substances are called neuromodulators. It is obvious that neuromodulators may not be expected to carry punctual neuron-to-neuron information at synapses as neurotransmitters do, but to exert modulatory influences over relatively large populations of cells. One thing is to speak close to somebody's ear; quite another is to speak the same words to a group of 30 people in a classroom. Norepinephrine and dopamine-containing axonal terminals are very often found amidst neural groups with receptors; in those cases, the substances act not as real neurotransmitters but rather as modulators. A variety of peptides, including some acting on opioid receptors and others with important neurotrophic function (like BDNF, an acronym for the brain-derived neurotrophic factor, which stimulates the growth of recently stimulated synapses), some substances with cannabinoid properties, some peptides with actions at receptors to opiates, and some hormone-releasing factors, are among the main neuromodulators. Several of these substances have been attributed a role in learning and memory processes [1, 8].

2.8   The Forgetting of Short- and Long-Term Memory

As described earlier, working memory is forgotten very rapidly over a few seconds without leaving a trace, which makes it the shortest form of memory. There are, as also mentioned, several varieties of short-term memory that last somewhere between a few minutes and a few hours and also leave no trace [10, 15, 18–22]. In mammals, these varieties of short-term memory are made by specialized neurons in the hippocampus and entorhinal cortex [20, 21] independently of the making and storage of the same information in long-term memory [20, 21, 23].

Short-term memories serve a similar purpose as living transitorily in a hotel while our house is being built. It maintains recently acquired information available while the brain builds the long-term memory store of it. It may be blocked by many treatments given into the hippocampus and/or the entorhinal cortex without altering the concomitant long-term memory, which shows that both require different and separate neuronal mechanisms in those two brain structures [20, 21]. The dichotomy between the mechanisms of short- and long-term memory in the brain is seen in widely different species through different mechanisms [20–22], which suggests that it represents an important evolutionary principle. The cellular consolidation of long-term memories, which once was thought to take place in a few minutes [14], actually requires several hours in which protracted protein synthesis-mediated events take place in the hippocampus and elsewhere (amygdala, entorhinal and posterior parietal cortex) [23]. After this, and over many days or even months, systems consolidation mechanisms elsewhere in the cortex take over and may add to or subtract from memory contents [2, 16].

Losing memories after short periods (minutes, hours) is something we learn to live with and is only perceived as forgetting by the part of the population that lives in constant fear of getting Alzheimer's disease, which after a certain age is quite large. "Doctor, I forget where I place my keys. Is this the beginning of Alzheimer's?"

The symptoms of brain degenerative disorders are usually much worse than simply forgetting where one puts the keys every now and then. Benign forgettings such as these are usually the consequence of distraction: our head is busy with other things while we are putting the keys here or there. Alzheimer's and other demented patients forget the faces of people who matter a lot to them (their sons and daughters, their wife) or forget their major skills and professions. If the short-term forgetting becomes disturbing, repetitive, or incapacitating, the person in doubt should consult a neurologist. The symptoms of dementia are not a consequence of the abundance of information sources in modern life; they occur in spite of this. They are a disease of old age and are rare nowadays in persons below 60 or 65. However, Alois Alzheimer described for the first time the disease that bears his name in a 51-year-old woman in 1906; people got older much earlier in those days.

Some people believe that the vast number of information we are nowadays bombarded with every day is detrimental to our memory capacity. Actually, most members of the health care professions understand that the wealth of information sources in today's life is a boon rather than the cause of disease. It is certainly more practical and less tiring to keep a good computer, library, and phone at hand than to try to keep all the information those peripherals carry in our own heads. Addresses, world news, football results, professional tips, medication doses, our schedule of appointments for next Monday, and bibliographical citations are better and more safely stored in electronic or written agendas than in our own distractible cerebral cortex. The cerebral cortex must be unimpeded and at hand for other, most pressing matters, like surviving, thinking, loving, and understanding reality.

2.9   Memory and Emotions

All memories are acquired and consolidated in some emotional state. All humans, perhaps all mammals, are at all times in a given emotional state, often difficult to describe; non-emotional moments are unknown to us. We may be more or less excited, more or less depressed or anxious, more or less enthusiastic or sleepy, and more or less happy or sad, but we are at all times in some kind of emotional state. There is no "unemotional" moment in humans and perhaps in other animals, mainly in mammals.

The memories most emotionally arousing at the time of consolidation are usually those best remembered. If at the time of consolidation the emotion is too big, however, so as to constitute something that should rather be called anxiety or stress, consolidation can be irreversibly impaired; the corticoids characteristically released by the adrenal glands during stress navigate the bloodstream and reach the hippocampus, where they block noradrenergic and other synapses involved in the consolidation process [17, 24].

We all remember where we were, what we were doing, and whom we were with when we saw on TV the second plane hitting the second tower on the ominous September 11, 2001. Those of us old enough to remember will never forget what we were doing and who we were with the moment we heard on the radio that President Kennedy had been shot. Nobody remembers the face of the person who sold us the tickets the last time we went to the movies, sometimes not even if the movie we saw was especially good. Nobody recalls the face of the attendant whom we asked to "fill it up with regular" last week or that of the cashier at the supermarket who gave us change last March 15th. The emotional impact of 9–11 or of Kennedy's assassination was much bigger than the two latter events and caused in us at the time a high emotional arousal. Jim McGaugh and Larry Cahill demonstrated a few years ago that the degree of emotional arousal, as processed by the basolateral amygdaloid nucleus, determines the accuracy and persistence of memories. Incidents accompanied by a small arousal are less memorable than those accompanied or "underlined" by a higher degree of arousal [9, 17, 25].

Probably the main reason why emotional arousal below the level at which it can be called stress enhances memory consolidation is the stimulation of the brain fiber systems that release the neurotransmitters, norepinephrine, and dopamine onto the hippocampal and amygdalar neurons that are in charge of memory consolidation [17, 24–26]. The cell bodies of the dopamine- and noradrenaline-producing neurons are in the mesencephalon and on the floor of the fourth ventricle, respectively, and both sets of neurons innervate the hippocampus, amygdala, and entorhinal and prefrontal cortex rather profusely [1]. At low-to-moderate doses, glucocorticoids share and enhance the actions of norepinephrine on consolidation [17, 26]. Both dopamine and norepinephrine stimulate a postsynaptic enzyme that mediates the synthesis of a cyclic derivative of the energy-laden compound ATP (adenosine triphosphate); its cyclic derivative cAMP activates an enzyme called protein kinase A, which phosphorylates substrates involved in glutamatergic transmission and in DNA transcription to mRNA [1]. Dopamine and norepinephrine thus promote DNA stimulation and protein synthesis in the hippocampus and amygdala [24], thereby enhancing their functions in the memory consolidation of memories just acquired. It has been known for over 40 years that hippocampal and amygdalar protein synthesis is necessary for the generation of long-term memories [8, 9, 17, 24].

Excessive arousal, which characterizes stress, blocks consolidation. Stress has been known for decades to be accompanied by large releases of corticoids from the adrenal glands; high levels of corticosteroids (in humans, cortisol) inhibit the brain noradrenergic systems involved in consolidation [26, 27]. Actually, this is typical of most drugs that enhance consolidation at low-to-moderate doses, including norepinephrine and dopamine [1, 17]: at low-to-moderate doses, they enhance consolidation, and at high doses, they inhibit it.

In non-biological terms, this has been interpreted as a safety valve by which the brain can cancel the recording of memories that can be too stressful. The occurrence of "blanks" at the time of retrieval (that seemingly insurmountable blockade of retrieval that may occur in students at oral examinations or in overanxious singers or lecturers at public presentations) is indeed explainable by the hypersecretion of corticoids caused by the stress of the public presentation, which act by inhibiting noradrenergic hippocampal or other brain mechanisms involved in retrieval [27].

It is hard, however, to assign a protective role to the failure of a student at an oral examination or of a singer at an audition or in a concert.

Interestingly, a low-to-moderate noradrenergic and dopaminergic stimulation of the hippocampus and prefrontal or parietal cortex, or a low-to-moderate release of corticosteroids, also stimulates retrieval [1, 28]; therefore, its neurohumoral modulation is very similar to that of consolidation.

In addition to glucocorticoids, many other hormones also related to general excitation or stress (peripheral epinephrine, the adrenocorticotropic hormone, vasopressin, oxytocin, β-endorphin, etc.) can influence consolidation, often by influencing brain norepinephrine or dopamine release or action [1, 17, 23, 24].

The modulation of the making and retrieval of memories by the biological correlates of emotional arousal and stress is a major chapter in the understanding of the art of forgetting. While we cannot voluntarily act on the biochemical effects of hormones, neurotransmitters, or modulators in the biochemical mechanisms of memory, we can train ourselves to regulate the expression of our emotions. Singers and students learn to control their fears and therefore stress at the time of a public presentation, and most famous artists and singers acquired that ability over a long time. Therein lies an art, both at the time of memory consolidation and at the time of retrieval, whereby forgetting may be curtailed or at least modulated. People can be taught to refrain from the expression of anxiety and thus eventually reduce anxiety itself or some of its consequences, like excessive corticosteroid release. The famous stiff upper lip may have a physiological function after all.

2.10   More on Memory and Emotions: Endogenous State Dependency

As said, there is no moment in the life of humans and perhaps most mammals that is unaccompanied by some degree of emotion, however small it may be. At all times we are more or less happy, sad, sleepy, aroused, anxious or calm, enjoying the situation or not, and contented or discontented. Therefore, when we acquire or consolidate any memory, it is always against a background or in the context of some kind of emotion, weak or strong. It is not the same to watch a movie about death when we are calm and happy or when we are anxious and depressed. We have seen above the evidence that emotional arousal, through noradrenergic stimulation of the hippocampus and the amygdala, leads to better and stronger memory formation and retrieval [9, 17, 23, 24].

As discussed above, many of the influences of emotion on memory formation and retrieval are attributable to the release of peripheral hormones [9, 17]. So are, indeed, the influences of emotion on heart rate, blood pressure, and mood. The ancient belief that memories were made by the heart (from which the phrase "learning by heart" comes from) is derived from the observation and feeling that tachycardia accompanies highly emotional memories, i.e., those that leave better memories [17, 25]. In many cases, perhaps mostly when the duration of the state of the brain and body brought about by emotions is relatively long or intense, the state itself may become incorporated to the cognitive cocktail that accompanies and modulate memory consolidation, retrieval, or both. So, when we learn and consolidate a fearful experience, we may also learn and consolidate the fact that brain norepinephrine and dopamine have been released, that peripheral epinephrine was also released, and that as a result of all this and of behavioral arousal the basolateral amygdala has been stimulated [9, 17, 23, 25]. When the neurohumoral and hormonal state that accompanies consolidation is repeated at the time of retrieval, this becomes usually better, more precise and more detailed. Sometimes retrieval is only possible when the animals are in the same state as they were at the time of consolidation; this is typical of thirst, fear, and stress. We do not remember how fear is unless we feel it. The reliance of retrieval on the state prevalent at the time of consolidation is called state dependency [29]. It was first described for exogenously administered drugs: when animals are trained in a given drug-induced state, say, under the effect of alcohol or opioids, they retrieve that memory well only when that drug state is repeated [29]. The first publication on this is over a century old and fictional: Robert Louis Stevenson's 1886 classic *The Strange Case of Dr. Jekyll and Mr. Hyde* [30], in which a concoction prepared by the doctor transforms him into a monster. The second was perhaps Chaplin's 1931 movie *City Lights*, in which the tramp played by Chaplin was recognized as a friend by a millionaire only when drunk; the first time he had seen the trump he was drunk. The state can be secondary to feelings or to the endogenous neurohumoral changes that accompany feelings: Dorothy Fields' title of the song she wrote with Jimmy McHugh, *I'm in the Mood for Love* (1935), reflects this.

So do the behaviors of people under the influence of sexual recollections and the accompanying release of hormones and the behavior of people under thirst: they tend to recollect memories relevant to sex and water drinking, respectively, over other memories. These are examples of endogenous state dependency [31, 32], by which memories made in a given hormonal and neurohumoral state, like those of fear, thirst, and sex, are better and preferentially retrieved when the individual is in the neurohumoral and hormonal state pertinent to fear, thirst, and sex, respectively.

Possibly memories of well-being, pleasure, and elation belong to this class too, i.e. to a state in which there is dopaminergic release in the so-called "brain pleasure areas" (mainly the nucleus accumbens, and a few others).

Thus, many biologically significant memories remain most of the time in a latent form, including those of recognition, aggressiveness, fear, sex, and thirst. We simply ignore them unless we are exposed again to the neurohumoral states that first brought them about. They remain lurking somewhere in the brain, in other cortical regions, or wherever else they are stored [2, 33], until the appropriate neurohumoral state is established and they can be retrieved [32, 33].

Possibly the drug-seeking behavior that characterizes addiction is the consequence of a form of state dependency initially induced by a drug that then becomes an endogenous set of stimuli.

As will be seen later, a prominent form to induce forgetting is by disuse. It relies on the eventually physical disappearance of synapses when they cease to be stimulated altogether. It was first described in detail in the synapse between the phrenic nerve and the diaphragm muscle and later in many other synapses from invertebrates to humans. Abstinence followed by prolonged disuse (i.e., in Alcoholic Anonymous or Narcotic Anonymous groups) is widely regarded as basically the best treatment for addicts. They are encouraged to discuss their addiction problems with, and follow the example of, fellow sufferers who have successfully cultivated abstinence and disuse. This is of course very difficult for many and can truly be viewed as an art. But it is certainly possible and usually the only way out for a large percentage of those who practice it.

Endogenous state dependency enables us to live in society more or less adequately, without performing out-of-context behaviors (like an erection or a cry for water while performing in a string quartet), and to respond appropriately to fear, sex, and thirst among other things. It was postulated by Zornetzer [31], demonstrated in laboratory animals by me and my group [32], and applied to clinical situations by Colpaert and his collaborators [33].

2.11   The Use and Disuse of Synapses

As mentioned earlier on, the long-lasting interruption of the use of synapses causes their atrophy and eventual disappearance. An excellent account of this is found in the classic 1964 book *The Physiology of Synapses*, by the great Australian neuroscientist Sir John Carew Eccles (1903–1997), a disciple of Sherrington, on experiments performed by himself and by others [34]. These experiments explored the effects of the drastic but very effective procedure of cutting the phrenic nerve, the one whose bilateral rhythmic activation by the brain's respiratory center stimulates the diaphragm every so many seconds and keeps us breathing. At the distal end of the sectioned nerve, its axons degenerate, and the various components of the muscle's postsynaptic apparatus, rich in receptors for acetylcholine (the neurotransmitter of synapses between motor nerves and voluntary muscles), soon disappear until no trace of them can be found.

The method of not doing something ever again is intrinsic to much of the type of education prevalent all over the world ("you must never do this again!").

Conversely, the repeated stimulation of synapses such as that between the phrenic nerve and the diaphragm [34] maintains the synapses functional. Indeed, it is common knowledge that the repeated reading of a paragraph or the repeated listening to a song or execution of a melody in the piano or recitation of a poem or a prayer maintains and strengthens their memory. There is in fact no way of learning the tables of multiplication, poems, or songs, let alone skills than through repetition. Andrés Segovia or Paco de Lucía played guitar as well as they did because they practiced each piece hundreds of times—so did Pavarotti and Sinatra with every song they sang or Herbert von Karajan with every Beethoven symphony he conducted.

As the experiments on the denervated diaphragm dramatically show, the disuse of synapses is a guaranteed way of making the information transfer through them disappear altogether. Most of the true forgettings that occur in us and in all animals are believed to result from the disuse of synapses, inasmuch as these are believed to be the depositories of memories, as Ramón y Cajal postulated in 1893 [3] and modern Neuroscience has shown.

Programmed neuronal death occurs normally during the development of nervous systems and is called apoptosis [35]. It is quite prominent in humans at the age when we evolve from the quadrupedal to the bipedal way of standing and walking (in general, 10–14 months of age). We thus effectively forget how to behave as quadrupeds because the cells that ruled the rapid crawling on hands and legs that we mastered at the age of 8 or 9 months have disappeared and turn into walkers for the rest of our lives, capable of relearning how to crawl only quite awkwardly because we have no neurons left which know how to really do it. We erased the systems of our brain that knew how to crawl shortly after we became 9 or 10 months old, by apoptosis.

Apoptosis is the best explanation of how we lose all neuronal representations of skills, people, animals, facts, or things that crossed our way just once in our lives and were unimportant to us and so were never repeated.

2.12   Failure of Memory Persistence as a Form of Forgetting?

Memories that last 1 day or more are called long-term memories. Some of them, however, last just for 3 or 4 days, and others persist for weeks, months, or years. Those that last years, like an old person's reminiscences of his (her) childhood, are often called remote.

In part, the persistence of memories is due to their emotional content and/or individual significance. As mentioned above, we remember for many years memories acquired under a strong emotion with a lot of arousal involved, a process that is usually attributed to enhanced gene activation and protein synthesis in memory-making cells in the hippocampus because of input from the basolateral amygdala and of noradrenergic and dopaminergic stimulation, that stimulates biochemical processes leading to enhanced DNA transcription and translation [1, 2, 23, 24]. But we also remember for many years events and facts whose emotional content is weak: some law of Physics or geometry learned in elementary school, a trivial scene from a movie, an isolated phrase or fragment of a song that by itself means nothing, and the face of somebody whom we have not seen for decades and whose name we have forgotten. There has been much hypothesizing, but no demonstration, that such trivial facts or episodes symbolize something of importance for our lives. We all know people who in high school or college studied just enough to pass a given test or exam and then forgot whatever they had learned forever, while some of his or her classmates studied the same item and remembered it for the rest of their lives.

A key factor in the persistence of these not-very-arousing memories could well be their successful reconsolidation shortly after the original learning (see next section).

Some authors have reported in the last decade on the existence of mechanisms in the hippocampus that regulate the persistence of simple memories acquired several hours before. All of these mechanisms are hippocampal; some are mediated by dopaminergic fibers, others by noradrenergic afferents or cholinergic synapses [8, 36, 37]. Some possibly act by promoting the release of BDNF in the hippocampus [36], others by other biochemical variables [37]. Their overall effect is to somehow make memories previously consolidated last longer: 7 or more days instead of just 1 or 2. This has been studied in laboratory tests in rodents and by field tests in humans [36]. It seems likely, as has been argued, that these purported mechanisms are simple prolongations of the action of those involved in memory modulation at the time of cellular consolidation [17].

Both in laboratory rats and in humans, the persistence mechanism(s) begins to fail in the second half of the life span: 1 year of age in rats and 40 years of age in humans. For example, rats can retain simple aversive memories very well for about a year but remember about 50 % at the age of 2 years, when they are close to the end of their life span. People below the age of 40 can recall for at least 1 or 2 weeks newly learned data on a remote sports event they had never heard of before, or incidentally acquired data on a movie they watched some days ago, but people above that age can retain such data for only 1 or 2 days [36]. A dopamine/norepinephrine enhancer given after the original learning can prolong the memory of people over 40 to the levels seen in the younger ones.

A question is "is such a treatment really useful or necessary?" To begin with, dopamine/norepinephrine enhancers, like all drugs, may have side effects. Then, perhaps the forgetting of incidentally acquired not-very-important memories from the last few days serves a useful purpose for those over 40. It is better for us to forget where we left our car in the parking lot of our office building 2 or more days ago so as not to confuse it with where we parked it today, which is what really matters when we leave the office tonight. It is better to forget trivial details of unimportant learnings acquired a few days ago in order to better remember those that may matter to us today (something our boss said at work, where we placed that file or that book or that screwdriver we need now the last time we used them, etc.). Perhaps this capacity for forgetting recent details in order to concentrate on "the big picture" or on what we are doing now is one of the reasons why executive tasks and positions (professor, chairman, director, general, coach) are more usually given to people those over 40 than to the snappier, more memorious but distractible younger colleagues [38]. Part of the "experience" that counts as an asset at a big job is perhaps the acquired ability to forsake trivial or useless details.

In "Funes the Memorious," Borges [6] said "he had the impression that Funes was not very good at thinking; to be able to think, we must forget in order to make generalizations."

2.13   Reconsolidation

It is now believed that memories persist in great part because of the recently discovered process of reconsolidation [39].

This is one of the two major possible consequences of retrieval without the reinforcement, the other being extinction (see Sect. 2.2).

If memories are retrieved in the first few hours or days after training, they may be enhanced or strengthened. The process can be altogether blocked by ribosomal protein synthesis inhibitors given systemically or, in mammals, into the hippocampus or the amygdala [38, 40, 41]. Reconsolidation seems to be a very general property of memories, and it has been described in many species, from crabs [41] to humans [42]. The most important experiments so far, in mechanistic terms, have been on rats [38, 40, 43]. At training-retrieval intervals longer than a few days, extinction predominates. The lack of early retrieval of most memories (e.g., those acquired yesterday that were not very important) probably accounts for their usual forgetting, which led Jim McGaugh to state 40 years ago that "forgetting is the most salient aspect of memory" (see Chap 1). It has been common knowledge for centuries that repetition is the time-honored (and certainly time-dependent) method of making memories last. We all know that there is no better way to learn poetry, songs, musical instruments, and the tables of multiplication than by repetition at relatively short intervals.

Extinction itself can be reconsolidated. This property of extinction permits a strengthening of the now classic psychotherapeutic use of it in exposure which basically consists of the extinction of unwanted memories, particularly those of fear [43, 44] (see Sect. 2.26, further on).

2.14   Practicing Memory

Real forgetting, as said, usually results from the disuse or disappearance of nerve cells and/or its synapses. It is not an art. Our will is not involved directly in the death of synapses or of whole neurons. Conversely, in order to maintain memory, one needs to exercise it regularly; this requires our will and that is an art. As discussed above, use enhances and disuse atrophies synaptic function. Memories are made at synapses and maintained by them; obviously, the more we put those synapses to work and keep them fit, the better memories will be made and maintained. This is a truism for all the functions carried out by synapses; the more we use the neuromuscular synapses of the fingers, the better it will be for the practice of guitar playing or for pecking on keyboards to write or to play music.

The practice of memory is the only way to keep its synapses in good shape, and the best exercise for that is reading. We do not know precisely what synapses are used in the making of each memory; but we know relatively well where they are in the brain: in the hippocampus, basolateral amygdala, entorhinal cortex, the prefrontal cortex, and, for habits, in the cerebellum and basal ganglia (see above). Within the cortex, it is obvious that visual memories will use the visual cortex and the pathways that link the retina with that cortex. Memories that link visual cues with other information will use, in addition, associative cortical areas. Verbal memories will use the regions related to language in the frontal, temporal, and parietal cortex. Memories with a motor component will use the motor cortex memories with an auditory component will use the auditory system, and so on. Obviously, complex memories that require the integrative action of different sensory modalities and/or their interaction with other elements will use different combinations of cortical and other regions.

One type of activity that requires the concurrent activity of all these regions is plain reading. This activity sets us apart from all the other animals. Many of them can articulate sounds that may be used for communication, i.e., what can be described as a primitive form of language, for example, in dolphins or whales. Real language involving grammar and syntax and reading are beyond their capacity and, as far as we know, are typically and exclusively human.

Humans vocalize sounds or syllables and use them to communicate, like dolphins or whales do, up to about the age of 1 year. In the second year of life, we begin to build and use words with one or a few syllables, which enable us to clarify and qualitatively enhance our communication power. Between the age of two and a half and three and a half years, we become able to make little phrases, with which we are in possession of a modicum of, that precious treasure, language. From then on, we translate everything we perceive or do into that new instrument and make memories in a symbolic way, not referring to other humans, animals, or things themselves but to the word or group of words they are represented by. This is obviously the beginning of an increasingly symbolic mode of cognition, in which things are not remembered as they really are but rather as they are called. We are suddenly in possession of a more sophisticated mind, and each day we become more capable of using the unbounded logic function of representations by combining letters into words and words into phrases. No wonder we become increasingly unable with age to remember things and concepts learned in our prelinguistic infancy; it is as we had learned them using another code, which is what we really did and is what immigrants who arrive in a new country early in life always do. At the age of 3 or 4 years, we are all immigrants in the new world of language.

Like other animals that have no languages at their disposal, as children we learned those prelinguistic items as unnamed components of reality, in a code without language. This is supposedly one of the main reasons, if not the main reason, why the repertoire of memories of humans begins with those we acquired after the age of 3 or 4, and very few of us can really remember anything from the years before that and, if so, very fuzzily or altogether indescribably.

A further step, usually accomplished between the age of three and a half to 6 years, is that of putting language into writing and being able to decipher it, i.e., of reading. Writing and reading not only further separate us from other animals but also differentiate us from those humans that do not know how to read and relegates them in all societies to those with the lowest share of power of decision. Gaskell said, "the race is not to the swift or to the strong, but to the wise," a phrase repeated often in class by my professor of physiology in Medical School, Bernardo Houssay, the first Latin American ever to win a Science Nobel Prize. To be able to read is the minimum of wisdom that is needed, not to win a race but at least to participate and have a say in the daily life of our information-dependent planet.

Reading is perhaps the only activity that practices and uses all the brain regions and memory forms described above. When we read we put into action working, verbal, visual, image, and semantic memory, often together with other sensory memories, as well as in many cases motor memory: our vocal chords are often stimulated by the retrieval of words, albeit usually subliminally.

The best possible recommendation to exercise the practice of memory and therefore support its maintenance is to read, read, and read. Evidently one can mobilize other forms of memory doing other things, but none of them mobilizes so many memory types and as thoroughly as reading or, for those with a visual deficiency, hearing together with touching and smelling.

Suppose our brain decides to read this section, "Practicing Memory." It will begin by the first letter in the title, a capital "P." When we read the letter "P," our brain instantly scans its vast archives searching for all the words it knows that begin with a capital "P": Peter, Pearl, Paris, Pantheon, Pennsylvania, Paraguay, etc. Most of us in the West read from left to right. Accustomed to this, immediately after detecting the P, our brain will carry out immediately a search to the right, and a few milliseconds later, it will find the letter "r." It immediately deletes Peter, Pearl, Paris, Pantheon, Pennsylvania, and Paraguay and initiates another search to the right, now looking for words in any language that begin with "Pr": Professor, Prince, Prometheus, etc. Except for Prometheus most other words are common nouns; if they have a capital P, they must be at the beginning of a phrase. The brain "looks" around and realizes that this is, yes, the beginning of a phrase, a title, actually. So the brain, right away, deletes the previous scans and makes a new one for words that can initiate a phrase in a book on forgetting and begin with "Pr...": Practice, Private, Principal, Probable.... By looking to the right, the brain finds that the letter that follows after "Pr" is an "a." The brain is left with few options, so it decides to risk an answer, which is something our brain does very often. It ventures "Practice" and finds that the letter that follows after the "a" is a "c." It goes on looking and finds a "t." The brain has now fewer choices: "Practice, Practical" among them. The brain looks right again, sees an "i," and ventures: "Practicing." A quick certifying look will tell the brain that it is the right word. Now a new look to the right, the brain finds a space and realizes that "practicing" was the first word; so it proceeds on to the first letter of the second word, which is an "m," and so on. Take note that each jump to the next letter and the corresponding scans that follow take a few milliseconds. We read quite fast indeed, and normally it takes a literate adult like us less than 1 s to read the entire title.

During that second, our brain used its various memory systems and performed a large exploration of its records of different words probably in several languages. And we have also seen again for an exceedingly brief time the faces of Peters and Pearls and felt a bit of the nostalgic aura we associate with Paris and the Pantheon of Rome ("the loveliness of Paris, the splendor that was Rome"), maybe reached out to memories of Tony Bennett and San Francisco and the early 1960s, all in a matter of a fraction of 1 s. Our brain unconsciously made a relatively large number of associations in a very short period of time. That is why reading is the single activity that thoroughly exercises more different memories and pathways and persons and concepts and songs and feelings in the brain than any other. It explores many things it knows from the most varied sources and often reminisces about places, faces, and even songs. To read is to exercise all sorts of memories and combinations of memories. No other intellectual exercise is as thorough and therefore as useful for the practice of memory as reading.

Borges became blind when he was about 50, and from then on till the end of his life 36 years later, he asked people to read to him in a loud voice, so he "read" through them, so to say. Several other major poets in history who also were blind had done the same, including John Milton. People who lose their eyesight usually do this. Many blind people nowadays use computers that can "read," i.e., that scan a text and spell out through a loudspeaker what the text says. People who are born blind may use these devices, and all people who cannot see may learn to "read" by touch in the Braille alphabet. Humans suffer comparatively more with the loss of eyesight than other animals, like dogs or rats that normally use their olfaction and audition to deal with the world. We humans are predominantly "visual" animals, and our visual cortex is larger and has more cells than that of other mammals [1]. Birds and insects have a sharper vision than we have, but they cannot read or write. Like somebody said, "nobody is perfect."

The example given above of how our brain reads included mention of the interaction between memories brought about by reading. For a similar analysis of reading but in the Portuguese language, see [1]. There is an additional level of interaction that takes place in everyday life, no doubt exercises memories, and may occur all the time and at any time without necessarily involving reading. The following example may suffice as a description of how this other level works. Suppose we see or hear a bird, and by simple association it brings to us the memory of a tree; when we see the tree, we remember fruits, and among the fruits we remember oranges and then a particular orange, one that was peeled by our mother for us when we were very little. So a bird ends up, in seconds, bringing to me the memory of my mother, the woman who first fed me and held me tight and whose relation to me permeated my entire life including the affection I feel for many people, animals, things, and places. No doubt the omnipresence of associativity is involved. All memories are considered to be associative one way or the other. Even the memory of habituation has been described as resulting from the association of a given stimulus with the lack of another stimulus.

But note that each of the associations triggered by the sight or the sound of a bird in the preceding paragraph was itself dictated by our memories: we remember that birds are associated with trees, which are associated with oranges, etc. Life and our general perception of it involve many experiences, and all of them may leave memories, all of which are associative. No wonder we forget so many; associations fail so much.

2.15 Reading, Knowledge, Physical Exercise, Social Life, Memory, and Illness

Many young people, especially high school students getting close to college age, often wonder: why study? And before them, children in elementary school when interminable classes try to teach them things they feel are useless, limited as they are to a reduced universe of home, school, the houses of friends and relatives, and the streets that communicate them with each other. What is the use of studying? It does not serve directly to play, to connect with others, to go anywhere that matters, to detect the novelties and dangers of the street, and to give or receive love. But mankind has been recommending for centuries to study, and the people that study more create better societies to live in. All this confronts children and especially teenagers with deep dilemmas that by the lack of experience, including study, they are usually quite unable to solve.

True, studying serves only to learn things and thus to expand the universe we live in, but that is a very big "only." It is certainly different to be bound in a nutshell where we can think of ourselves as kings of heavens and earth, like Hamlet, than to know that there are many countries, languages, cultures, climates, and things that we will gain by *understanding*. The word *only* means not just the expansion of our universe through knowledge; its significance becomes better understood when we realize that dementias, such as that of Alzheimer's disease, can make us lose that expansion and "throw us back into being a nobody," like some people with that disease told me, one of them my father and another one a world-famous physicist.

The importance of reading, learning, and thereby making memories is further illustrated by the numerous studies showing that the decline of memory that usually accompanies old age is much reduced and begins later in people who have read a lot and studied a lot, like actors, actresses, teachers, and professors. So does the loss of memory that occurs in Alzheimer's and other forms of dementia. Many studies in large urban populations have consistently shown that. Memory losses in the aged and particularly in the demented aged correlate not so much with the socioeconomic status, stress (some Holocaust victims lived to a very old age with their minds well active), or sadness (many depressed old people live a long and mentally active, although not a happy life). Memory losses in the aged are lower in professions that require reading. Actors, actresses, singers, musicians, teachers, and professors, who have to read a lot because of their lines of work, usually remain mentally active and very fit in terms of memory up to a much higher age than those who chose a less fortunate or demanding profession.

Other major factors in preventing memory loss in old age are physical exercise, possibly in part because of the general improvement in general circulation including brain blood flow and perhaps in part by the stimulation of the formation of new neurons (see further on). Another factor in preventing memory loss is a healthy balanced diet, not only because of the prevention of high blood levels of "bad" cholesterol and adult diabetes but also to maintain adequate brain nutrition. Still another factor is awareness of the life around us; the news of our neighborhood, city, country, and world; and, last but not least, social life. An excellent account of these factors can be found in the Web site of Harvard Health Publications on Preventing Memory Loss.

The usually benign memory loss of humans above the age of 60 or 70 is characterized more by slowness in making or retrieving memories than by an actual significant loss of their content. It does include a partial loss of short-term memory, which fades faster. Older people experience the famous "tip of the tongue" phenomenon much more often than young people (The director of *Vertigo*? Ah, his name is at the tip of my tongue. English, wasn't he? Just give me a few minutes and I'll remember it").

When the failure of short-term memory and the difficulty in remembering older declarative memories above the age of 60 or 70 is intense, it is called mild cognitive impairment (MCI) and is said by some to affect over 20 % of those over 75. The decision by the psychiatrist or psychologist of whether the person being examined has an MCI or falls short of it is given by the response to the simple questions about general orientation (place, date), short-term memory, and other tasks in the so-called mini-mental state examination (MMSE), a questionnaire that can be answered in a few minutes. MCI patients typically have difficulty in remembering the names of people or of new places they have visited and sometimes may have a degree of trouble remembering the flow of a conversation or lose attention for short spells, together with an increased tendency to misplace things or to forget addresses, phone numbers, or passwords, particularly if newly acquired. In many cases, individuals are aware of these difficulties and overcome most of them with increased reliance on notes and calendars. Those affected typically cannot remember right away something that is being asked from them but will spontaneously come out with the correct response a few seconds or minutes later, occasionally hours or days later. A relatively large portion of those with MCI (over 50 % according to some accounts) proceed to Alzheimer's or some other form of dementia. Some authors consider MCI as an early, prodromal phase of dementia, but most do not agree. Anyway, whether with MCI or early dementia or none of the two, many aged people with mild memory impairments go about performing their activities and lead a basically normal, productive, and emotionally satisfactory life for years.

Medical evaluation of people with suspected MCI includes the mentioned MMSE plus a thorough exploration of the memory complaints, the duration of the problem, and whether there are other concomitant cognitive complaints (problems with organization, planning, etc.). The physician should be aware of the patient's medical history, the medications taken, and the possibility of depression, which is often accompanied by such symptoms but in which, once it is treated with psychotherapy and drugs, the symptoms readily subside. Anxiety disorders may also seriously hamper cognition and are also treated well with psychotherapy and appropriate drugs. Sleep disorders are also a common source of memory trouble and can also be treated medically. When dealing with patients with a history of superior mental power, physicians and health caretakers should always remember that a well-trained brain can do things that untrained brains cannot even dream of doing [45, 46] and is often able to fend off even the earlier phases of Alzheimer's disease for a while, some more, some less.

As to tricks or, rather, strategies to overcome the symptoms of MCI, the Argentine Nobel-Prize winner for Physiology mentioned in an earlier section was famous for his use of little written notes on people he had to see in the incoming days, into which he jotted details on his or her doings and those of the family and which he carried in his pockets. He rapidly studied them before meeting those people and enjoyed surprising them by showing what appeared to be a fantastic memory for his age. He began doing this at about the age of 50, when he felt his memory was becoming not so good but was still far away from any symptom of MCI. Indeed he did not show any sign of MCI when he was over 80 and still very active at work. He died with no signs of dementia or any other brain disorder, after he had to stop working for the first time in his life for a few months after a fall.

My father, a university professor who had been hospitalized with Alzheimer's disease for a few years, 1 day was asked by his former graduate students to go with them to a scientific meeting and present a poster on some work he had done with them. He studied a lot during several days, went over and over the old data many times, and stayed very quiet throughout, always in the hospital. The day of the poster presentation, he put on his best blue suit and finest red tie. His former students stayed close to him during the session just in case, but they had no need to break in or interrupt him. My father's poster presentation was impeccable.

2.16   The Art of Forgetting: A Second Approach

What can we do to develop an art of forgetting in the midst of the flood of information that continuously impinges on our working memory system? How can we discriminate in that flood between signals and noise? How can we detect what really matters, keep it, and throw away the rest?

As mentioned in an earlier section, no less than one of the fathers of Neuroscience, Santiago Ramón y Cajal, felt at a loss when confronted with such questions back in the 1920s, when the inflow of information was a fraction of what it is today and he was over 80. The response to those questions is that we can learn how to cope with all that by practice. Air traffic controllers learn how to do their job by sheer practice, and they very rarely fail. Air traffic is much more complex today than it was a few years ago, and it works perfectly well all of the time in all airports all over the world. Life-saving tasks and procedures are performed correctly and successfully by many people in the middle of war or catastrophe situations. Physical scientists perform perfectly well even in very complex situations that require entangled reasoning and calculations and so do top-notched athletes. Circus artists, racing car drivers, and fighter plane pilots learn to perform well and avoid life risks even in the most dangerous circumstances. In all these examples, the working memory of the performers learns how to discriminate between signal and noise by numerous trial-and-error computations short of the errors that could be dangerous, both when they were learning and while performing what they learned. Actually, they never stop learning because each day they learn a little more and that matters.

The procedure is similar to the one we apply every time we are inadvertently close to a car accident or a serious mistake. In the nick of time, our working memory tells the rest of the body to turn right or to jump out and away from whatever we were into. Our working memory or "central executive of information" applies the same type of reasoning when it tells us not to pay attention to the color of the shirt of the truck driver whom we are about to collide with or that of the eyes of the person who is about to shoot us. Just turn right or jump away. When we are operating an artery, we must quickly differentiate it from the tissue around it and from another artery. In order to detect the fading cry of a survivor in the middle of the rubble of a building after a bomb or an earthquake, we must make our mind oblivious to all the noise around us, no matter how loud. Trained and attentive people can do this; that is why so much training is required to be a good driver, a soldier, a surgeon, or a fireman. Mothers are the champions of the art: they learn how to detect the faint cry of their baby three rooms away on the first trial.

For working memory, the art of forgetting consists of acquiring the ability of discriminating against the unnecessary, the accessory, against the noise and in favor of the signals.

In other forms of memory, especially long-term memories, the art of forgetting consists of the avoidance of activity in the neural pathways of undesired memories. This is most often difficult, and we need the help of psychotherapists to learn it, except for the memories of drug addiction which can be satisfactorily erased by the interaction with fellow sufferers, as commented above. The art of forgetting can be imitated by falsification or by the brain-induced inhibition of remembering or retrieval, which usually functions satisfactorily enough. These processes will be commented upon in the next sections. They serve to cope with the persistence of unwanted memories, which eventually haunt us and forbid us to carry on a normal life; the worst example of such "persecutions" by bad memories is the dramatic psychiatric syndrome called post-traumatic stress disorder (PTSD).

2.17   Denial and Memory Falsification

A certain degree of denial and memory falsification are inherent to life and necessary for it to go on. Denial and falsification are most often relatively automatic and unconscious. We just set aside some unwanted memories as if they never happened (*denial*) or change them into different memories if it suits us better (*falsification*). The problem with these is, as the Germans say, *Lügen haben kurze Beine* (*lies have short legs*). Sooner or later, unforeseen stimuli or circumstances may bring those bad memories back in their ugly original real form in a fully unexpected moment, and that may cause us serious trouble. For example, if a frightening or humiliating episode we have denied flashes into our mind in the middle of a risky airplane maneuver, we might fall; if it happens in the middle of the performance of a violin solo, we may be booed by the public.

We know very little about the mechanisms of *denial*. It has not been specifically studied so as to correlate it with regional brain changes in blood flow in preventing memory loss flow, a technique used to indirectly estimate the putative involvement of this or that brain area that is involved with some particular influence on behavior. Denial is considered to be related to mechanisms of anxiety, but there is no proof of this. Some have postulated that it may be linked to mechanisms of memory repression, a term coined by Sigmund Freud to indicate the cover-up or suppression of a given memory (see a section ahead); but this is even farther away from proof.

Denial is a much used form of passive coping particularly dangerous in people who have diseases they prefer not to have and that can lead them to death, like heart conditions, obesity, diabetes, or the addictions. It leads them away from treatment and thus endangers their life. People that claim not to have diseases, syndromes, or problems that to everybody else are obvious are often referred to as "being on denial."

More is known about *memory falsification*, which is the modification of existing memories with false data carried out voluntarily or involuntarily. Nothing can really be said about voluntary memory falsification; it is simply forging a lie and eventually believing in it.

Concerning the involuntary or spontaneous making of false memories, they have been studied in detail by the American psychologist Daniel Schacter. He wrote an excellent book called "The seven sins of memory" [47]; one of the seven sins is falsifying memories. Schacter uses the word "sin" in a figurative or metaphoric sense, of course. Indeed, perhaps the word "sin" is itself always metaphoric.

Another American psychologist, Elizabeth Loftus, studied this issue for years and made an experiment that became famous [48]. She showed a film clip of a car accident to a number of students and then asked them details of what they had seen. She subdivided the subjects into several groups; the only difference between them was one word in the questions asked to each group. Some were asked "About how fast were the cars going when they hit each other?" Others were asked the same question substituting the word "hit" by "smashed," "collided," or "bumped." Those asked the question with the word "smashed" reported the highest speeds, followed in descending order by those that were asked the question with the words "collided," "bumped," and "hit." Some of the students were also asked "Was there any broken glass in the film?" A large majority of those submitted to the "hit" question correctly answered "no." A large majority of those exposed to the "smashed" question said "yes" [48]. Clearly, just one word in the post hoc question made a big difference in what the subjects actually remembered.

In the decades following this experiment, Loftus was asked many times to explain and expose flaws in eyewitness testimony. Her observations and ideas are recognized all over the world and have led to changes in the legal system of the USA and other countries. She had clearly showed for the first time that exposure to one different word at the time of questioning can change the account of a fact by eyewitnesses. Marcia Chaves and I asked students to study a 16-line text with factual information on the 1954 soccer world cup, which took place before any of them had been born. One or three hours after studying the text, some of the students received a slip of paper saying that sports commentators of the time said that cup was the worst ever. Two days later we asked the subjects ten specific factual questions on the material they had studied: results, winners, incidents, etc. Those who were given the slip of paper saying that the cup was bad scored a mean of 3 correct responses over 10. Those who did not receive the slip of paper or received another slip of paper saying that the cup was good scored about 7. The slip of paper saying that the cup was bad had no effect if presented 6 h after studying the text. Thus, an unspecific qualitative post hoc comment shortly after learning significantly affected factual information acquired earlier [49]. This endorses and enhances Loftus's findings, applying it to written material rather than to events acquired visually and to time-locked post-training qualitative changes referent to the material previously read rather than to changes of words at the time of testing, perhaps related to early memory consolidation [17]. Indeed, long-standing evidence shows that pharmacological or behavioral manipulations performed shortly after training are more effective for altering memory than those performed at other times [9, 17], a finding which we have discussed in previous sections and on which actually the entire idea of post-training (cellular) memory consolidation relies.

The most famous contemporary storyteller in the Spanish language, Nobel-Prize winner Gabriel García Marquez (who died while this book was being written), wrote a fantastic autobiography (*Vivir para contarla, Living to tell the tale*) [50] containing a large amount of obviously false data and says in its opening phrase that "(Our) life is not the one we lived but the one we remember, and (it is important) how we remember it in order to tell the story" [50]. That book is a pleasure to read.

Making false memories is not necessarily a sin. Coming to believe the false story about ourselves can be. Of course the possibility of creating false memories by the purposeful interpolation of false information can have ethic or legal connotations. But aside from that or from García Marquez's literary use of them, false memories are regularly made by normal people (see below) and are in many ways part of our daily life. Most of us have mental images of the room we had when we were six as very large and of our neighborhood as paradise (or hell) which are usually quite false. The world is full of heroic statues of persons who were in fact bad people, often mounted on a rearing horse and holding a sword up with their right hand while looking at some illuminated distant spot. But then, our books of history are also full of false information about people whom our nations undeservingly elevated to the category of great statesmen or heroes. The cult of such figures reminds me of Clint Eastwood's famous line in *Unforgiven* (1992), "Deserving has nothing to do with this." Nations seem to need the cult of heroes and often do not really investigate who some of these heroes really were and whether they really deserved bronze statues or would better be forgotten. All the big countries I know with the interesting exception of Brazil cultivate the memory of military heroes, bronze statues, and all. In Brazil there is nothing comparable to the cult of Washington, Bolívar, or San Martín in the Americas, or of the generals of the two World Wars in the USA, or the Cid in Spain, or Jeanne d'Arc in France.

Old people are prone to create false memories, usually changing the stories, the deeds, or the subjects of members of their family or friends. Indeed, old people are considered by many psychologists to privilege happy or at least lively memories over sad ones. My mother, when she became old, often "remembered" funny or extravagant things her younger brother had done and attributed them to me: "remember, Ivan, when you drove your Studebaker to the fountain in the middle of the city square?" When my uncle supposedly did that (if he ever did), I was 3 or 4 years old, and when I learned how to drive, Studebakers were hard to find in our streets anymore. I have never driven one. Children between 5 and 10 years old also sometimes tell tall stories, which they rigorously come to believe, in which they had defeated some monster or outlaw or otherwise acted in defense of their little sister or their mother. Girls that age (more than boys) not only have imaginary friends but then remember as true some of the things they did with them or said to them. Most of the false memories created by young children or old people are rather unconsciously concocted, involuntarily made, and quite innocent. The same cannot be said of the false stories invented by psychopaths in order to serve their evil devices, and these must not be confused with false memories. Part of the strategy of psychopaths when caught may consist in simulating that the stories they invented were false memories.

Some simple-minded and suggestible adults, most often with little formal education, can lead themselves to believe lots of stories if they honestly feel it is to their benefit to do so. Some descendants of European immigrants in the Americas or Australia often purchase grossly retouched photographs of people dressed like in the time of their grandparents, then repeatedly tell themselves and their neighbors that those were really their grandparents, and end up after years of such simulation believing the story themselves or at least become not very sure of whether that is true or not. A person I knew, descendant from Germans, once saw me in the street after I had just returned from some trip to Germany. I asked him from what part of Germany were his parents from; he looked at me blankly and with tears in his eyes told me he really did not know. I saw him again several years later, and his general aspect, assurance, and wide smile showed right away that something good and major had happened in his life. "You know," he said, "I discovered after some research that my parents were from Pomerania," and beamed at me. A common friend, days later, told me that this was not true but that after seeing some movie my acquaintance came out of it convinced that his parents were Pomeranians and would not discuss that with anyone anymore. Obviously, our now better-dressed and beaming common acquaintance had self-implanted a false memory of where in Germany his family came from and required now no blank looks and no tears in his eyes to remember them or talk about them. He had discovered a homeland.

Memories invented by psychopaths may actually be "planted" into suggestible persons in order to serve the purposes of the one who invented the story. For example, it became sort of a fad in California 20 or 30 years ago for young people to accuse one of their parents of sexual abuse. These were suggestible and usually ignorant young people who were induced by unethical psychotherapists to believe that, so they could send their parents to trial and get from them money which they would divide with the therapists. The scam was destroyed in most cases with the help of well-known psychiatrists and psychologists summoned as expert witnesses by defense attorneys. In some cases the liars had their way, and the reputation of the accused parents was seriously damaged. There are similar stories in many countries, and many in movies, of similar scams including false evidence and sometimes the making of false memories in relation to illegal possession of drugs planted by law-enforcing officers or others in order to put someone in jail.

False memories imply a form of forgetting. Real memories have to be deleted totally or in part in order to replace them with a false memory. If innocent, false memories can help some people have a better life, like my acquaintance that came to believe he descended from Pomeranians. But in many other cases, it may have harmful, damaging, or dangerous consequences. False memories are mostly created spontaneously and in response not to our will but often to deep desires, which in turn involve more false memories or incorrect data. The suppression of real memories that has to take place when they are substituted by false ones is often a voluntary act.

In most cases, false memories are not built on purpose; they are made by the brain on its own, without the resource to consciousness. The brain makes use of real or imagined data, including in the latter things our brain wished that had happened openly or secretly. They are often the result of wishful thinking or desired but not-really-true components. The use of false memories may modify the past we think we had, and some credulous people may genuinely think that they had really been musketeers or cowboys or surgeons or something else. It is sometimes difficult to extricate the real from the false in people who have and use many false memories. That is why there are so many successful liars around.

2.18   Forgetting by Large Populations

Surveys carried out in recent years show that a majority of Brazilian and British citizens do not know whom they voted for in the last elections. In 2002 most Americans could not locate Afghanistan in the map, in spite of the fact that most newspapers had shown where it was many times in their front pages in recent years. In 2003 a majority of Americans did not know where Iraq was, in spite of the two wars their country had already fought there in the recent past. A large number of Germans cannot remember who started World War I, and a large number of Austrians do not know that Hitler was born in their country. Russians learn in school that their country had a war with Japan in the early 1900s and lost; most adult Russians cannot remember their country ever had this war. A large number of young Argentines and Brazilians do not know and cannot believe their countries were under military dictatorships in recent decades, often including the year in which they were born. I saw an American war veteran visiting the Anne Frank museum in Amsterdam in the

1980s with his teenage children; these did not remember what war their father had fought in and were bored by the museum, "Oh Dad, let's go somewhere else, this is boring, we don't like it." Their father had fought against Hitler in the Second World War.

Part of that forgetting is due to the fact that the corresponding information was not well consolidated or repeated enough. Part was due to the fact that the people to whom the questions were asked would rather forget what the questions were about. But its result in terms of the history of their own countries, including the recent history, was that their insertion into the history of mankind in the recent decades was not or no longer a part of them. Norberto Bobbio [1] said that "we are what we remember," to which one may add, "and we are also what we chose to forget." There is no way to explain the advent of Neo-Nazism in the world after 1945, with all the dangers that this advent entails other than as a consequence of the choice of large populations to forget their recent history.

Another part of that forgetting is due to the relentless effort of dictatorial governments, using all the means of communication at their disposal for years, to erase all news and all mention to their deeds. I have seen people who lived near concentration camps in Germany who bona fide did not know that the camps existed (or, perhaps more believably, repressed that knowledge because they had been thoroughly trained to do so for years and it embarrassed them in front of foreigners like me). This explanation, however, does not apply to the forgetting by the democratic population of Brazil or England of whom they voted for in the last election or to that of Americans of where Afghanistan or Iraq are, at a time when the sons of some of their neighbors were deployed in those countries. These forgettings are all best explained by a generalized lack of attention or of interest, with the resulting fragility and evanescence of the corresponding memories. The full practice of democracy requires obviously much better than that.

Be it ignorance, planted disinformation, or both (Germany, Brazil, and Argentina have experienced long sequences of the two), the generalized forgetting of major historical events is an omen for our future as free individuals. The art of getting rid of it will require a much bigger effort than that already made by democratic societies.

2.19 Surviving Through Neuronal Death

Programmed cell death or apoptosis is a regular component of normal cell turnover and is less widespread in the brain probably because most neurons do not reproduce [1, 35]. It involves a series of well-defined biochemical mechanisms and is modulated by hormones and by development [35]. Neuronal hyperactivity, as seen in the excessive activation seen in epilepsy or in experiments using prolonged stimulation with the excitatory transmitter, glutamate, may lead to apoptosis. Indeed, neuronal death usually is the hallmark of disease. But it does play a role in the genesis and recovery of a variety of brain conditions. In an earlier section we discussed its role in the passage from the age of quadrupedal to that of bipedal behavior, when we are

10–14 months old. In that case, there is an unconscious and inherent art in the programming by the brain of what cells must go and of which ones are to stay and in the selection of what motor behavioral information is to persist and what part of it must be discarded [1].

Epileptic brain tissue results from inherent and somewhat ill-defined metabolic processes in some neurons that arise from metabolic derangements of intrinsic properties of neurons, or as a consequence of irritative effects of low irrigation (ischemia) like in cerebrovascular disease, or of tumors or other lesions. In most cases, epilepsies are treated and well controlled by a number of drugs. In others, there is resource to surgical procedures.

Atrophy and sclerosis of the hippocampus and/or the rest of the tip of the temporal lobe leading to overstimulation of hippocampal cells are very often followed by a form of epilepsy that is highly recurrent and incapacitating, inasmuch as that region is crucial for the formation and retrieval of declarative memories. Temporal lobe epilepsy is regularly treated by surgery and restores temporal lobe function to normal, particularly memory formation and retrieval.

In some areas of the brain, apoptosis is followed by the multiplication of neurons near the place where their counterparts have died; the cerebellar cortex and the hippocampus are among those areas. The production of new neurons is called neurogenesis. Nobody has studied this in more detail than Fernando Nottebohm [49] in singing birds, which lose and rebuild large neuronal population every mating season.

2.20   The Famous Case of Patient H.M.

Temporal lobe surgery for the elimination of local epileptic foci was in its primordial stages in 1953 when a well-known American neurosurgeon, William Scoville, extirpated the frontal portion of both temporal lobes in a patient called H.M., who died in 2008 [52]. Since his death, most medical and scientific writers have decided to call him by his own full name, Henry Molaison. We will, however, go on referring to him as H.M., because his real name rings fewer bells than his clinical name, which made him famous in the study of human memory. H.M. had epileptic foci in both temporal tips, and his epilepsy was very intense and did not respond well to medications. But in 1953 the surgery of temporal lobe tip focal epilepsy was new, a great distance from what it is today, when hundreds of temporal lobe operations to treat focal epilepsy are performed weekly in many specialized centers all over the world, with little or no side effects.

The occurrence of seizures in H.M. fell drastically after the surgery, but so did his ability to form new declarative memories, and he lost all memories formed in the weeks prior to the operation. He did retain the capacity to learn a few manual or perceptive skills, like some puzzles including one called "the Tower of Hanoi," which was more or less popular 30 or 40 years ago, but was unable to describe how he completed them ("Funny, I was trying to figure it out, but I couldn't"). He had great difficulty in learning the way from his room to the toilet or the face of someone whom he had been talking with a few hours before. He did remember tunes older than 1953 but could not remember tunes heard the day before and which he had liked. He remembered general information of world events prior to 1953 but was unable to learn who were John Kennedy or Pope John Paul II or how the Korean war ended.

H.M. once described his own condition quite graphically, saying "My mind is like a sieve." Information stayed in it for a few minutes but were not retained any longer.

For many years, when there was no computerized tomography or magnetic resonance, it was assumed that H.M.'s surgical lesion consisted of bilateral ablation of the hippocampus, and in fact this belief spurred the idea that the hippocampus played a key role in memory. H.M. was submitted to a careful and very complete neuropsychological study by Brenda Milner that set the basis for the study of what we know today about the role of the hippocampus in memory [52]. In the 1990s, when he was 66, finally H.M. was examined by magnetic resonance imaging by Suzanne Corkin and her associates [53], who found that the lesion was quite symmetrical and included the temporal polar cortex, most of the amygdaloid complex and the entorhinal cortex, but only about two thirds of the hippocampus; its caudal-most part was intact on both sides, although atrophic. Therefore, the effects of the lesion are most likely mostly the result of the almost complete bilateral entorhinal cortex ablation. The entorhinal cortex is the recipient of most of the afferent and efferent connections of the hippocampus, however [54], so its bilateral ablation can be viewed as a bilateral isolation of the hippocampus from the rest of the brain.

H.M. died in 2008, but his legacy to the biological understanding of memory will remain with us for a very long time [55]. In 2014, Suzanne Corkin and her coworkers published a sophisticated postmortem analysis of H.M.'s brain based on histological sections and 3D reconstruction [56]. Several other patients with bilateral hippocampal or temporal lobe lesions that included the hippocampus have been studied over the years [57, 58]; the common trait between all of them and H.M. is the deep anterograde and retrograde amnesia, especially for episodic memory: they cannot make new episodic memories (anterograde amnesia) or remember those acquired during the weeks or months prior to the lesion (retrograde amnesia). One patient with supposedly "pure" hippocampal lesions has been described as having an amnesia restricted to episodic (autobiographic) memories, sparing to a very large extent those of the semantic type (facts, rules, concepts) [57]. Many reproductions of these clinical lesions in laboratory animals have been made, and they have all been found to cause memory disturbances of the same kind, that is, declarative memories, particularly those of the episodic type are affected, and procedural knowledge is spared.

Patients with selective bilateral lesions of the amygdaloid nucleus are relatively rare. They have difficulties both in remembering and in acquiring the emotional component of episodes but do learn and remember their cognitive components. From a series of slides on a young boy hit by a car and taken to an emergency ward, these patients remember the non-emotional components of the story as well as normal volunteers, but they fail in the more emotional aspects and seem unperturbed by having to talk about them [59].

2.21   When Forgetting Is Not an Art: Amnesic Patients

In many people forgetting is due to pathology. The pathological loss of memories is called amnesia. As said, the incapacity to form new memories is called anterograde amnesia. The loss of preexisting memories is called retrograde amnesia.

There is no art in this; memories are simply lost because the brain is incapable of retaining them, as H.M. said. Anterograde amnesia is due to a deficient functioning of neurons or to the lack of enough synapses in good working order to process new memories. Retrograde amnesia for memories lost in the first day or two after they were acquired is due to the fact they were weak and did not recruit enough synapses and/or enough protein synthesis to make them last. Retrograde amnesia for longer-lasting memories is usually due to an impaired function or to the death of neurons or atrophy of synapses secondary to different diseases, many of them degenerative. The previous section discussed just one case, H.M., whose amnesia was due to bilateral extirpation of parts of his temporal lobes. But there are diseases that can cause amnesia sometimes even more incapacitating than that of H.M.

The main groups of diseases that affect memory are the dementias (see page 5). Among these, the one with the highest prevalence is Alzheimer's disease affecting about 20–25 % of people over the age of 80. In recent years, with the increased life expectancy of humans all over the world, people begin to age later, and the diseases of old age also begin later. The age of onset of Alzheimer's disease has moved forward; as mentioned elsewhere, the first case of this disease described by Alois Alzheimer was a 51-year-old woman. In the 1980s at least 30 % of those over 60 were considered to have the disease. The second most common type of dementias are those of vascular origin secondary to hindrance of arterial blood flow by clots and/or cholesterol or fatty material deposits on atherosclerotic plaques in patients with high blood pressure; the picture is often accompanied after some time by thickening of the walls of the blood vessels. The dementias due to impairments of brain blood irrigation may occur by themselves (15–20 % of all dementias) or, more often, be superimposed on Alzheimer's lesions (about another 15–20 % of all dementias). The memory loss of dementias is progressive over several years, and at the end of its cycle, it can include the faces of people we know well, places that used to be very familiar, and even language.

In an earlier section, mild cognitive impairment (MCI) was mentioned as a syndrome affecting perhaps as much as 10–20 % of those over 60. It consists of a difficulty in remembering all sorts of memory, particularly episodic memory. When pronounced, MCI is viewed by some as a prelude or prodromal stage of Alzheimer's disease or of some other form of dementia.

The amnesia of neurodegenerative diseases is treated symptomatically by a number of drugs that act on the brain cholinergic system, which is a major modulator of memory consolidation and retrieval acting on the hippocampus and other forebrain sites; the best known of these drugs are rivastigmine, galantamine, and donepezil. Much evidence points to a failure of that system in the dementias; transgenic mice lacking a functional forebrain cholinergic system show profound amnesia and some other neurological disturbances [60]. The intracerebral or systemic administration of a blocker of one of the receptor types to acetylcholine, scopolamine, has been known for many years to induce a very deep transient amnesia and was used as a model ever since [61].

Brain glutamatergic transmission underlies the functioning of most of the brain systems involved in the consolidation and retrieval of declarative memories [23, 24]. Drugs called ampakines that enhance regular glutamatergic transmission by an action on one of the glutamate receptor types in the hippocampus and elsewhere are also used with a degree of success in the treatment of the amnesic component of dementias.

In their early phases, Alzheimer's patients may preserve "islands" of good memory. It has been postulated that they represent real "islands" of relatively undamaged tissue in the entorhinal or parietal cortex that functions more or less normally. I knew two university professors (one of them a Nobel-Prize winner) who, in some years into their respective Alzheimer's disease, remained still capable of carrying on good dialogues with former students who came to visit them. Some of these students told me that these dialogues were usually quite fruitful, and in both cases, the old professors were able to exchange useful and correct bibliographical or other scientific information with them—to be sure, often in the middle of senseless tirades. One of them told me that their brain-damaged professor was able to give them the full reference (names of authors, initials, journal title, year, volume, first and last page) to significant articles they had discussed together and he knew by heart. Another student told me that his professor gave him important and very sensible practical advice on his career even while already committed to the bed where he would die a few days later. There often are "islands" of well-preserved memories among the flood of broken-down or lost memories in the brains of Alzheimer's patients. It is possible that some of those "islands" correspond to "islands" of intact or more or less intact brain matter.

So one must not dismiss Alzheimer's patients as people who are already "out" because in some of them, some pieces of their minds might work beautifully until near the end. Nobody is out in the fight for life until he or she dies.

Depression, a disease that at one time or another affects about 6 % of the population (more women than men, in a proportion of over 2 to 1), is probably not accompanied by any easily detectable anatomical change (some have described an atrophy of the hippocampus in depression but many have denied it) and is very often accompanied by a degree of amnesia. Unlike that of the degenerative diseases, which is largely irreversible, the memory impairment of depression usually reverses upon treatment of the original disease, which is carried out usually by a combination of cognitive psychotherapy and antidepressant drugs. The last 5 or 10 years have witnessed an increase in the number of antidepressant drugs. They enhance the action of the brain neurotransmitters norepinephrine, dopamine, and serotonin, which are involved in the regulation of mood and are suspected to fail in clinical depression. The antidepressants more used these days are those that selectively enhance the action of serotonin: citalopram, escitalopram, sertraline, and fluoxetine. Also used are amitriptyline, bupropion, and others that preferentially increase noradrenergic or dopaminergic transmission.

Depression must not be confused with sadness. Unlike the latter, it does not pass with time or consolation and involves a variety of symptoms besides sadness—prominently, sleep disorders, anhedonia (the incapacity to feel pleasure), a hypersecretion of corticosteroids, and a variety of autonomic and central nervous disturbances. It usually comes together with a degree of anxiety—which is related to an increased corticosteroid secretion—and, if untreated, may lead to suicide. It is, in fact, the most frequent cause of suicide. Thus, it is a serious psychiatric disease that requires treatment with both psychotherapy and antidepressants. It is irresponsible to treat it just with one or the other; it is a potentially very dangerous disease, and all means available should be used to treat it. Pseudo-ideological ready-made phrases that result from sheer ignorance are inapplicable. ("I don't take drugs." "I don't like psychiatrists or psychologists." "This is a mental disease and should be treated with words.") Drugs *and* words treat mental diseases; both are specifically designed for that. Depression often comes as part of a mixed syndrome or disease in which the patient oscillates between depression and mania or enhancement: the bipolar disorder. Depression is three times more frequent in women than in men and more frequent in teenagers and in the aged, two periods of life in which sweeping changes in behavior and in one's relation to the world occur, often beyond our control.

In children or very young people, sadness may not be a prominent symptom of depression. The therapist must be careful and search for the other major symptoms of the disease before establishing a diagnosis.

In the elderly, depression may be confused with early dementia, in which case it may be called "pseudo-dementia." In some old patients, the early phases of dementia may actually be associated with a depressive state, because the patient feels that "his or her mind (or brain) is going." The differential diagnosis is often difficult but must be done, because the outcome of each of these two diseases is very different. Depression can be treated successfully, whereas dementia just goes on. The two outcomes require very different steps by the patient and his or her family in their preparation for the future, both emotionally and economically.

It is counterproductive to treat the amnesic component of depression by itself leaving the rest of the depressive disorder untreated. The sudden reappearance of a bad memory that was thought to be forgotten may trigger a suicidal episode in a depressed individual. Depression is a very serious disease and should be treated as a whole, so the amnesia will fade away with the rest of the disease once psychotherapy and antidepressant medication have taken hold.

As said, some patients with Alzheimer's disease or other dementias may at some point realize that "their mind is going." This may lead them, usually early in the course of their disease, into a depression that needs to be treated as it would if their bearer was normal; but these are demented and often frail old patients, and the treatment is not easy: often the patients' deliberately or involuntary forget to take the medication for one or other ailment and may seriously compound their problem. Bette Davis once said that getting old is not for sissies. Getting depressed when being already demented is even worse; being a relative or an attendant of such patients also is.

Finally, a transient form of deep global amnesia may develop in people who experienced intense head trauma. It results from edema of the hippocampus and neighboring structures and lasts for as long as the involved cells remain swollen. It involves recent and remote memory and is called transient global amnesia and is treated with corticosteroids and other drugs that may alter the water metabolism of the brain, and it can be completely reversed. The older memories (those of years ago) reappear before those of months ago, then the memory of episodes or semantic knowledge acquired in the past few weeks return, and a retrograde amnesia of minutes or a few hours just preceding the trauma often remains [2]. Transient global amnesia may occur as a consequence of psychological rather than physical trauma; in those cases it is not accompanied by visible lesions of any part of the brain—just by fMRI signs of reduced hippocampal irrigation.

2.22   Anterograde Amnesia by an Interference with Consolidation

One of the major discoveries of Jim McGaugh, referred to above [15], is that memories are not acquired right away in their definitive form but take time to consolidate. This led him to propose that there is a process of brain consolidation after acquisition, on which hundreds of scientists worked in the following 60 or more years [23–25, 27]. Post-training cellular consolidation was thought initially to last only a few minutes [15], and therefore post-training treatments that could affect it (head trauma, anesthesia, electroconvulsive shock, or a variety of brain-depressant drugs) were thought to be able to act only during those few minutes; it was only much more recently that post-training consolidation and the time for treatments to act upon it were found to be of several hours, including the treatments that cause retrograde amnesia [23, 24]. Retrograde amnesia may sound hard to believe to a policeman arriving at the scene of an accident to whom one of the involved says "I have no idea how I turned up here." But it certainly is a major boon to all of us who once in a while have to suffer surgeries under general anesthesia.

McGaugh and then many others, notably Larry Cahill [25, 26], found that the norepinephrine β-receptor antagonist, propranolol, was among the drugs that induced retrograde amnesia. Many of its analogs shared this effect. Some years ago, propranolol was hailed by some as a drug that could erase "bad" memories, and they advocated its use for those purposes. It certainly does but only if given very shortly after a traumatic experience, in the phase when the laying down of memory traces is sensitive to and modulated by β-noradrenergic synapses in the amygdala and hippocampus [27]. But it is of little practical value, because people usually do not carry propranolol or other β-receptor blockers in their pockets to use only after a catastrophe of some sort takes place. Only people with the specific cardiovascular disturbances that require the use of β-antagonists as medication carry them in their pockets, and they don't want to forget about their heart trouble so when they reach the cardiologist, they have something to say.

It makes no sense to carry "forgetting" pills in the pocket, and it will still make no sense the day a pill like that is really discovered. This is unlikely to happen in the next decades outside movies and articles written by people untrained in the health professions.

2.23 Neuronal Branching and the Suppression of Branching, Neurogenesis, and Neuronal Death as Adaptive Phenomena

As mentioned in the first section, Ramón y Cajal postulated in 1893 that the biological basis of memories must be structural and thereby functional changes in synapses. In his day, Neuroscience was not even born; he was actually one of its founding fathers of it; in fact, he discovered synapses. He established modern neuroanatomy, which was the origin of Neuroscience.

An important form of learning in some birds is called imprinting. It takes place at a very precise time in early life and consists of learning to follow a moving figure that passes near them. Usually the figure is the mother, and chicks, little gooses, or little ducks learn to follow their mothers to wherever they go: water, a source of food, etc. One student of mine in São Paulo once bought a little duck in the street just before a class in Medical School and brought it to class. The little duck, once placed on the floor, walked after her and followed her wherever she went all day. It was imprinting time and she did not know.

Anna Katarina Braun and her coworkers in Germany discovered that the pruning of dendritic branches in a brain area roughly equivalent to a mammalian association cortex is fundamental for filial imprinting in the domestic chick. Later, the biochemical basis of this process was worked out in detail [62]. Imprinting is a form of learning that lasts for several days and is due to the suppression rather than to the growth of neuronal branches. As discussed in earlier sections, a similar process consisting of actual neuronal death (apoptosis) underlies the conversion of the quadrupedal model of brain we use below the age of 10–14 months to the bipedal model we will use from then on.

Another, important form of learning occurs in singing birds and is due to the generation of new nerve cells (neurogenesis). A massive production of new neurons takes place in the walls of the lateral ventricles of the brain of adult canaries and other birds in the season in which they sing to attract females. These neurons migrate to the appropriate nuclei that govern singing and last till the end of the singing season. Then they suffer apoptosis and are not replaced until the next period of singing, the following year. This extraordinary discovery was made in 1981 by Fernando Nottebohm [51] and changed overnight the concept prevalent until then that all animals are born with all the neurons they will ever have, i.e., that neurons did not reproduce and therefore adult neurogenesis did not exist. Nottebohm discovered adult neurogenesis, and his work helped to understand not only the seasonal character of bird singing but also neurogenesis in general.

Recent work by Fred Gage and Henriette van Praag in Bethesda, Maryland, has shown that physical exercise can be a powerful stimulant for neurogenesis in the hippocampus of aging mice [63]. Given the genetic and neuroanatomical analogies between human and mouse brain, and the increasing practice of exercise to maintain a good brain function in aged people, this result is of great potential medical interest. The practice of physical exercise to guarantee a better aging and an increased life span was initiated several years ago on the basis of serendipity: people who swam, walked, or ran often remained healthy until an older age. This was attributed vaguely to an enhancement of heart function and blood circulation all over the body and may also depend on brain neurogenesis.

Braun's and Nottebohm's findings show that the brain practices very sophisticated forms of the art of learning and of the art of forgetting by relatively coarse methods, consisting of the more or less wholesale production of new neuron or parts of neurons or in their pruning or death. These processes govern fundamental aspects of life, like whom birds must follow when little and when and to whom they must sing when grown-up or when we humans should drop crawling and rely on walking to places.

2.24   The Acquisition of New Memories

Obviously, if the consolidation of memories depends on synaptic alterations, but the synapses that are modified are not those activated directly by the sensory aspects of the experiences that lead to memories, perceptions are not changed by experiences. We see or hear the same before and after a learning experience, but our remembrance of that experience will change its meaning, which is what we call memory. Many experiences require the sensory organs: the eyes, the ear, the tongue, the nose, and the fingertips. But memories are not formed in them and cannot be formed in the neuronal prolongations that communicate the sensory organs with the brain, because those prolongations do not have synapses. Memories must be acquired at brain sites synaptically connected with, but outside, the sensory pathways; otherwise our perceptions would change. All the sensory pathways emit collateral branches that make synapse with a diffuse network of neurons that extends over the dorsal part of the mesencephalon and is called the mesencephalic reticular formation. The reticular formation projects massively to all the cerebral cortex and governs arousal. We wake up in the morning because the mesencephalic reticular formation, activated by collaterals from the auditory pathway, detects the sound of the alarm clock and projects to the cortex. Much of the sensory information processed by the mesencephalic reticular formation projects to integrative areas of the brain that respond to all sensory modalities and thus may serve interactive functions, like memory. The best known of these integrative areas are the hippocampus, the amygdala, the entorhinal cortex, and parts of the prefrontal, occipital, and posterior parietal cortex (Fig. 2.1). These areas are all secondarily activated by sensory experiences, in parallel to the sensory pathways, and participate in the processes of learning and making memories.

But memories are also made of interoceptive information (the contraction and relaxation of skeletal muscles, of intestinal contractions, and of insights or combinations of preexisting memories). Many of these internal sources of learning are unconscious, such as most muscle contractions and elongations or the many small changes in the acidity of the stomach. There are also memories with an important motor component, such as those involved in walking, writing, singing, or playing a musical instrument. The brain areas responsible for all these other components of memory are relatively well known, with the exception of the insights and eventually the religious or ethical feelings that may be parts of the insights that lead to some memories. Imaging studies, in particular fMRI (functional magnetic resonance imaging) on which lay people place so much mechanistic hope, are of little help except for the information they provide about blood flow in relatively large portions of the brain. They tell us merely what areas of the brain are presumably activated by the processing of memories, but they are uninformative as to the cellular/molecular processes involved or about the activity of the neurons that participate in the learning and memory processes or others.

It is also obvious that in order to acquire and consolidate memories, the brain, or at least its regions involved, must be in a minimum but sufficient state of alertness or arousal. Sleeping brains may detect stimuli and especially learned stimuli: there are many instances of mothers who slept through bombings but even in the midst of them woke up to the specific sound of their newborn crying. But evidence for real learning during sleep has been much searched for in the past 60 years and is still lacking.

Also, as analyzed in preceding sections, at least in humans and most probably in all mammals, the acquisition, consolidation, and retrieval of memories always take place in a given emotional state, big or small, weak or strong, happy or sad, intense or vanishing, and stable or fleeting. Among humans, only psychopaths really experience very little emotion, and that is why they can commit such heinous deeds as they do. But there are numerous proofs that even they are never completely unemotional. In fact, they are essentially great deceivers and can fake emotions very well.

Memories are more precise, carry more information, and persist longer when their hippocampal cellular consolidation takes place during a state of emotional arousal, as mentioned in a preceding Sect. 2.9. This is the consequence of two parallel events: activation of the basolateral amygdala, which is the main "sensor" of emotional arousal, and an activation of the brain noradrenergic and dopaminergic systems, which stimulates enzymes that indirectly regulate hippocampal and amygdalar protein synthesis [23, 24].

The biochemical pathways activated in the brain cells that make memories are many, and they are interwoven in a relatively complex fashion. There are more than 50 enzymatic systems interacting with one another to various degrees [17]. The interactions between these systems generate two physiological processes in the hippocampus and elsewhere which are believed to underlie the making of memories: long-term potentiation (LTP) and long-term depression (LTD) of the glutamatergic synapses involved in the consolidation process (see Sect. 2.5 and Figs. 2.1 and 2.4). The biochemical changes that have been described following acquisition (i.e., at the time of cellular memory consolidation, see above), particularly in the hippocampus and in the basolateral portion of the nucleus of the amygdala [23, 24], are remarkably similar to those observable in these structures during LTP or, in some cases, during LTD [23].

The number of biochemical pathways involved in LTP and LTD and/or in memory consolidation and the intricacy of their interactions [24] makes the search for the cells that acquire and consolidate memories quite difficult, and nobody who really works in this field reasonably hopes to ever pin down "the biochemical/ cellular map" of any given memory. A major source of complexity is the fact that the inter- and intracellular pathways involved in different memories interact with each other. How many times we begin learning or retrieving something about one person, thing, or situation and end up mixing that inextricably with data on other person(s), things, or situations? Somebody's face reminds us unwillingly and often unconsciously of the face of somebody else. It frankly makes no sense to search for the intra- or intercellular pathway of any single, isolated memory. In addition, it is very possible to understand many memories and memory types knowing as we do only a small fraction of their biochemical or neuronal map. We need to know only a percentage of the words in any text or oral collection of words in order to understand the sense of all the rest (see Sect. 2.14 on how the brain can make accurate guesses of letters or words it has not yet read). To determine the full molecular map of any given memory would be as wise as to determine the exact amount of sodium or of cockroach legs in the walls of the cathedral of Nôtre-Dame or any other church for that matter. It would tell us nothing significant about the history, beauty, or meaning of those buildings (see [17]).

2.25 Repression

Repression is a word coined by Sigmund Freud to describe the inhibition of the expression of some memories in order to keep them out of our conscious mind. It is usually practiced by the brain to obliterate painful, fearsome, or otherwise disagreeable or unwanted memories in order to carry on a regular life.

It can be voluntary or involuntary [64]. In both cases it has a major physiological function. If women did not repress their memories of the pain associated with delivering a baby, very few of them would have a second child. If patients did not repress memories of the pain and discomfort experienced at the dentist, they would all tend to die of an untreated dental condition of some sort. If all of us did not repress the problems caused by getting up on cold winter mornings, we would not finish school and remain illiterates until the final despondency reserved for us by a world that increasingly prizes knowledge.

For years, repression was "explained" only by psychoanalytic hypotheses and remained incomprehensible to neuroscientists. Also, many psychiatrists and psychologists used to deny the existence of voluntary repression. This was in spite of the fact that we have all experienced the successful practice of it. We all sometimes are able to erase even the slightest reminiscence of people or places we abhor, let alone their names. Recent neurophysiologically oriented research has now furnished solid evidence for the operation of definite brain systems in voluntary repression.

Three different sets of fMRI observations in humans show that different groups of neurons are involved in the voluntary repression of memory. In 2004, Michael Anderson, John Gabrieli, and their coworkers taught subjects pairs of words and then, when exposed to one of the words of each pair, they asked to repress the memory of the other member of the pair [64]. They found that the repression was accompanied by increased blood flow in the dorsolateral prefrontal cortex, which suggested that this area was activated, and by a reduction of blood flow in the hippocampus, which suggested that this structure was inhibited during repression. This suggested that there is an active repression process mediated by the anterolateral prefrontal cortex and inhibition of the hippocampus. The data of Anderson and his coworkers suggest that the same processes may be in charge of unconscious repression.

More recently, Benoit and Anderson using similar methods found an additional mechanism for voluntary repression, by which two other subdivisions of the prefrontal cortex interact so as to induce substitution of the unwanted memories by others [65].

A third group [66] also using fMRI techniques has disclosed the activity in voluntary repression of two more sets of brain structures in the frontal cortex independent of the former two.

Thus, now there are concrete neuroscientific explanations for the repression process, all based on fMRI measurements on voluntary repression. Perhaps the brain areas involved in the repression of memories of different content are also different; in any case, the evidence shows that several brain regions are involved in voluntary repression. Unfortunately, unconscious repression is not amenable to direct scientific exploration for obvious reasons. So, unless we accept the possibility that it should obey to more or less the same mechanism as the voluntary variety, it will remain an untestable construct, like most of the old psychoanalytic postulates. Psychoanalysis antedated modern Neuroscience by a number of years, and its postulates generally bear little if any relation to what we now know about brain functions after more than one century of Neuroscience.

2.26   A Therapeutic Use of the Art of Forgetting

Fear memories are usually traumatic and thus among the most unwanted memories imaginable. They are, however, important for survival. It is essential to remember that tigers bite unless next time we go to the zoo we do not place our fingers into their cage or to know that buses can hit us unless we cross by the pedestrian zone when the light is green for us. Clearly, the retrieval of fear memories out of context is not desirable; for example, to recall the 9–11 scenes or the angry look of a tiger when one is about to sleep or have sex or read a book is not desirable. The recall of fear memories when needed, at times in which it is important to remember strategies to escape from fearsome situations or to fight them off, is very desirable; it is, in fact, the biological reason why fear memories are so resilient and long-lasting. So, one thing is to keep fear memories for use at appropriate times; quite another is to retrieve them out of context and/or at inappropriate moments of our lives.

The most unwanted disturbance caused by traumatic fear memories is post-traumatic stress disorder (PTSD) in which a given episode or series of episodes (death threat, torture, sexual assault, serious injury, extreme danger) become impossible to get rid of, and the sufferer experiences recurring flashbacks of a "bad" memory, avoidance, and extreme anxiety for months after the event(s) at non-chosen times. PTSD is more common in women, very common in war veterans, and uncommon in children. It makes a regular life impossible or at least extremely difficult. It belongs to the category of anxiety disorders along with panic and phobias, of which it is by far the worst.

PTSD is regularly treated by extinction procedures in which the therapist exposes the patient to stimuli akin to those that were responsible for the original trauma repeatedly. This procedure is called exposure therapy and is not without its drawbacks, inasmuch as some patients cannot tolerate being exposed even to incomplete and clearly artificial versions of the original traumatic stimulus (e.g., pictures or talks instead of the real thing). But if performed judiciously by a trained therapist, it is the best procedure available and the only one that may result in a real cure.

Extinction therapy was actually introduced by Freud in the therapy of phobias back in the 1920s under the name of habituation, which as discussed above actually designates a different thing. But then, Freud never liked Pavlov very much.

Two different possible adjuncts to extinction have been recently proposed for exposure therapy. One is the surprise presentation of novel stimulus or set of stimuli at some moment during an extinction session of fear memory. By a mechanism involving the processing of the memory both of novelty and of extinction by hippocampal LTP or LTD, such a procedure has been found to significantly accelerate extinction in laboratory rats [66]. The other procedure that was found to accelerate extinction in rats is vagal stimulation during extinction, a relatively simple manipulation that has been approved for the US Food and Drug Administration for its use in depression and other psychiatric conditions [67].

2.27 Accessory Memory Devices: An Adjunct to the Art of Remembering and to the Art of Forgetting

We all use accessory memory devices external to us just like cell phones, tablets, and CDs are external to computers.

They comprise books, libraries, all sorts of written or electronic agendas, tablets, and telephones, plus people who know about us or who can tell us things we do not know. These "peripherals" help us a lot in our daily life, which today would be unconceivable without them.

When we leave them aside, we give our brains a rest, however, accustomed as we are to their use. That rest is often very much needed; we need to "disconnect" every now and then from these accessory sources of information. Lest we choose to live in a virtual world that is not exactly the one that surrounds us. The obvious example is not to use the cell phone while driving; another example is not to use our tablet or laptop computer when we are attending a lecture or going out with our girlfriend or boyfriend or our son when we have not seem him for a year.

Accessory memory sources have been accused by some as the villains of our cognition in modern times. But they are instead a great help to give our brain memory systems a break, especially those that participate in working memory and in the retrieval of semantic and episodic memory. It is by far safer, and strains the mindless, to look up a phone number or the name of the director or main actor of this or that film in a written or an electronic notebook rather than to rely on our own busy and, alas, so often not-very-reliable head. Computing devices are of course of no use for procedural memories; we do not use cell phones or tablets to tell us how to ride a bicycle. When the vehicles are more complex than a bike, such as a plane or a rocket, we certainly need them.

Crystal radios were an obvious first approximation to radio transmission that was bound not to last. Computer devices will certainly last much longer. It is by far more judicious to learn how to use them than to complain against them as Cajal did with the crystal radios of his neighbors. The use of computer devices while driving, listening to a lecture, or talking with a significant other is of course to be avoided, as is the reading of thrillers or singing the national anthem during those occasions. There is plenty of life outside computers, tablets, and cell phones. Most of it is outside, in fact.

References

1. Izquierdo I. Memória. Porto Alegre: Artmed; 2011.
2. Squire LR. Memory and brain. London: Oxford University Press; 1983.
3. Ramón y Cajal S. Neue Darstellung vom histologischen Bau des Zentralnervös System. Arch Anat Physiol (Anatomy). 1893; 419–72
4. Greenough WT. Morphological and molecular studies of synaptic memory mechanisms. In: Gold PE, Greenough WT, editors. Memory consolidation – essays in honor of James L. McGaugh – a time to remember. Washington: American Psychological Association; 2001. p. 59–77.
5. Geinisman Y. Structural synaptic modifications associated with hippocampal LTP and behavioral learning. Cereb Cortex. 2000;10:952–62.
6. Borges JL. Ficciones. Buenos Aires: Emecé; 1944.
7. Pavlov IP. Lectures on conditioned reflexes. Oxford: Oxford University Press; 1927.
8. Myskiw JC, Izquierdo I, Furini CRG. The modulation of fear extinction. Brain Res Bull. 2014;105:61–9.
9. McGaugh JL. Memory – a century of consolidation. Science. 2000;287:248–51.
10. Sutton MA, Carew TJ. Behavioral, cellular, and molecular analysis of memory in Aplysia. I: intermediate-term memory. Integr Comp Biol. 2002;42:725–35.
11. Goldman-Rakic P. Architecture of the prefrontal cortex and the central executive. Proc Natl Acad Sci U S A. 1995;769:71–83.
12. Fuster JM. Cortex and mind. Unifying cognition. New York: Oxford University Press; 2003.
13. Weinberger DR, Harrison P. Schizophrenia. 3rd ed. Hoboken: Wiley; 2011.

14. LeDoux JE. Coming to terms with fear. Proc Natl Acad Sci U S A. 2014;111:2871–8.
15. McGaugh JL. Time-dependent processes in memory storage. Science. 1966;153:1351–8.
16. McClelland JL, McNaughton BL, O'Reilly RC. Why there are complementary learning systems in the hippocampus and neocortex: insights from the successes and failures of connectionist models of learning and memory. Psychol Rev. 1995;102:419–57.
17. McGaugh JL. Making lasting memories: remembering the significant. Proc Natl Acad Sci U S A. 2013;110:10402–7.
18. Myskiw JC, Rossato JI, Bevilaqua LR, Medina JH, Izquierdo I, Cammarota M. On the participation of mTOR in recognition memory. Neurobiol Learn Mem. 2008;89:338–51.
19. Izquierdo LA, Barros DM, Vianna MRM, Coitinho A, de David e Silva T, Choi H, Moletta B, Medina JH, Izquierdo I. Molecular pharmacological dissection of short- and long-term memory. Cell Mol Neurobiol. 2002;22:269–87.
20. Izquierdo I, Barros DM, Mello e Souza T, Souza MM, Izquierdo LA, Medina JH. Mechanisms for memory types differ. Nature. 1998;393:635–6.
21. Izquierdo I, Medina JM, Vianna MRM, Izquierdo LA, Barros DM. Separate mechanisms for short- and long-term memory. Behav Brain Res. 1999;103:1–11.
22. Emptage NJ, Carew TJ. Long-term synaptic facilitation in the absence of short-term facilitation in Aplysia neurons. Science. 1993;262:253–6.
23. Izquierdo I, Bevilaqua LRM, Rossato JI, Bonini JS, Medina JH, Cammarota M. Different molecular cascades in different sites of the brain control consolidation. Trends Neurosci. 2006;29:496–505.
24. Izquierdo I, Medina JH. Memory formation: the sequence of biochemical events in the hippocampus and its connection to activity in other brain structures. Neurobiol Learn Mem. 1997;68:285–316.
25. Cahill LF, McGaugh JL. Mechanisms of emotional arousal and lasting declarative memory. Trends Neurosci. 1998;21:294–9.
26. Roozendaal B, McGaugh JL. Memory modulation. Behav Neurosci. 2011;125:797–824.
27. De Quervain DJ. Glucocorticoid-induced inhibition of memory retrieval: implications for posttraumatic stress disorder. Ann N Y Acad Sci. 2006;1071:216–20.
28. Barros DM, Mello e Souza T, De David T, Choi H, Aguzzoli A, Madche C, Ardenghi P, Medina JH, Izquierdo I. Simultaneous modulation of retrieval by dopaminergic D1, β-noradrenergic, serotoninergic1A and cholinergic muscarinic receptors in cortical structures of the rat. Behav Brain Res. 2001;124:1–7.
29. Overton DA. Basic mechanisms of state-dependent learning. Psychopharmacol Bull. 1978;14:67–8.
30. Stevenson RL. The strange case of Dr. Jekyll and Mr. Hyde. London: Longmans & Green; 1886.
31. Zornetzer SF. Neurotransmitter modulation and memory: a new neuropharmacological phrenology? In: Lipton MA, DiMascio A, Killam KF, editors. Psychopharmacology: a generation of progress. New York: Raven; 1978.
32. Izquierdo I. Endogenous state dependency: memory depends on the relation between the neurohumoral and hormonal states present after training and at the time of testing. In: Lynch G, McGaugh JL, Weinberger NM, editors. Neurobiology of learning and memory. New York: Guilford Press; 1984. p. 333–50.
33. Colpaert FC. State dependency as a mechanism of central nervous system drug action. NIDA Res Monogr. 1991;116:245–66.
34. Eccles JC. The physiology of synapses. Berlin: Springer; 1964.
35. Kuan CY, Roth KA, Flavell RA, Rakic P. Mechanisms of programmed cell death in the developing brain. Trends Neurosci. 2000;23:291.
36. Izquierdo I, Bevilaqua LR, Rossato JI, Lima RH, Medina JH, Cammarota M. Age-dependent and age-independent human memory persistence is enhanced by delayed posttraining methylphenidate administration. Proc Natl Acad Sci U S A. 2008;105:19504–7.
37. Parfitt GM, Barbosa AK, Campos RC, Koth AP, Barros DM. Moderate stress enhances memory persistence: are adrenergic mechanisms involved? Behav Neurosci. 2012;126:729–34.

38. Nader K, Schafe GE, LeDoux JE. Fear memories require protein synthesis in the amygdala for reconsolidation after retrieval. Nature. 2000;406:722–6.
39. Sara SJ. Strengthening the shaky trace through retrieval. Nat Rev Neurosci. 2000;1:212–3.
40. Milekic MH, Alberini CM. Temporally graded requirement for protein synthesis following memory reactivation. Neuron. 2002;36:521–5.
41. Carbó Tano M, Molina VA, Maldonado H, Pedreira ME. Memory consolidation and reconsolidation in an invertebrate model: the role of the GABAergic system. Neuroscience. 2009;158:387–401.
42. Forcato C, Rodríguez ML, Pedreira ME, Maldonado H. Reconsolidation in humans opens up declarative memory to the entrance of new information. Neurobiol Learn Mem. 2010;93:77–84.
43. Schiller D, Monfils MH, Raio CM, Johnson DC, LeDoux JE, Phelps EA. Preventing the return of fear in humans using reconsolidation update mechanisms. Nature. 2010;463:49–53.
44. Schiller D, Kanen JW, LeDoux JE, Monfils MH, Phelps EA. Extinction during reconsolidation of threat memory diminishes prefrontal cortex involvement. Proc Natl Acad Sci U S A. 2013;110:20040–5.
45. Patihis L, Frenda SJ, LePort AK, Petersen N, Nichols RM, Stark CE, McGaugh JL, Loftus EF. False memories in highly superior autobiographical memory individuals. Proc Natl Acad Sci U S A. 2013;110:20947–52.
46. LePort AK, Mattfeld AT, Dickinson-Anson H, Fallon JH, Stark CE, Kruggel F, Cahill L, McGaugh JL. Behavioral and neuroanatomical investigation of Highly Superior Autobiographical Memory (HSAM). Neurobiol Learn Mem. 2012;98:78–92.
47. Schacter D. The seven sins of memory. Boston: Houghton Mifflin Harcourt; 2001.
48. Loftus EF, Palmer JC. Reconstruction of auto-mobile destruction: an example of the interaction between language and behavior. J Verbal Learn Verbal Behav. 1974;13:585–9.
49. Izquierdo I, Chaves MLF. The effect of a non-factual posttraining negative comment on the recall of verbal information. J Psychiatr Res. 1988;22:165–70.
50. García Márquez G. Vivir para contarla. Buenos Aires: Sudamericana; 2002.
51. Nottebohm F. Neuronal replacement in adulthood. Ann N Y Acad Sci. 1985;457:143–61.
52. Scoville WB, Milner B. Loss of recent memory after bilateral hippocampal lesions. J Neurol Neurosurg Psychiatry. 1957;20:1–21.
53. Corkin S, Amaral DG, González RG, Johnson KA, Hyman BT. H. M.'s medial temporal lobe lesion: findings from magnetic resonance imaging. J Neurosci. 1997;17:3964–79.
54. Hyman BT, Van Hoesen GW, Damasio AR. Memory-related neural systems in Alzheimer's disease: an anatomic study. Neurology. 1990;40:1721–30.
55. Squire LR. The legacy of patient H.M. for neuroscience. Neuron. 2009;61:6–9.
56. Annese J, Schenker-Ahmed NM, Bartsch H, Maechler P, Sheh C, Thomas N, Kayano J, Ghatan A, Bresler N, Frosch MP, Klaming R, Corkin S. Postmortem examination of patient H.M.'s brain based on histological sectioning and digital 3D reconstruction. Nat Commun. 2014;5:3122. doi:10.1038/ncomms4122.
57. Rosenbaum RS, Köhler S, Schacter DL, Moscovitch M, Westmacott R, Black SE, Gao F, Tulving E. The case of K.C.: contributions of a memory-impaired person to memory theory. Neuropsychologia. 2005;43:989–1021.
58. Dede AJ, Wixted JT, Hopkins RO, Squire LR. Hippocampal damage impairs recognition memory broadly, affecting both parameters in two prominent models of memory. Proc Natl Acad Sci U S A. 2013;110:6577–82.
59. Squire LR. Memory and brain. New York: Oxford University Press; 1987.
60. Martyn A, De Jaeger X, Magalhaes AC, Kesarwani R, Gonçalves DF, Raulic S, Guzmán MS, Jackson MF, Izquierdo I, MacDonald JF, Prado MA, Prado VF. Elimination of the vesicular acetylcholine transporter in the forebrain causes hyperactivity and deficits in spatial memory and long-term potentiation. Proc Natl Acad Sci U S A. 2012;109:17651–6.
61. Deutsch JA. The cholinergic synapse and the site of memory. Science. 1971;174:788–94.
62. Bock J, Braun K. Filial imprinting in domestic chicks is associated with spine pruning in the associative area, dorsocaudal neostriatum. Eur J Neurosci. 1999;11:2566–70.

63. Van Praag H, Shubert T, Zhao C, Gage FH. Exercise enhances learning and hippocampal neurogenesis in aged mice. J Neurosci. 2005;25:8680–5.
64. Anderson MC, Ochsner KN, Kuhl B, Cooper J, Robertson E, Gabrieli SW, Glover GH, Gabrieli JD. Neural systems underlying the suppression of unwanted memories. Science. 2004;303:232–5.
65. Benoit RG, Anderson MC. Opposing mechanisms support the voluntary forgetting of unwanted memories. Neuron. 2012;76:450–60.
66. Depue BE, Curran T, Banich MT. Prefrontal regions orchestrate suppression of emotional memories via a two-phase process. Science. 2007;317:215–9.
67. Myskiw JC, Benetti F, Izquierdo I. Behavioral tagging of extinction learning. Proc Natl Acad Sci U S A. 2013;110:1071–6.

Summing Up

3

In the preceding pages several aspects and forms of the art of forgetting were discussed:

1. To forget quickly the items processed by our working memory is part of its function.
2. Memory falsification is often used by our brain as a defense mechanism.
3. Voluntary repression exists as a brain mechanism.
4. Extinction, habituation, and discrimination or differentiation are forms of learning that rely on the art of selectively cancelling responses that have become useless or undesirable. This may have therapeutic value in people or animals hurt by fear or otherwise disagreeable memories like phobias, panic, generalized anxiety, and post-traumatic stress disorder.
5. State dependency, in which the brain reserves its right to respond only when placed again in the neurohumoral state in which it was when it first learned something. State dependency is more an art of becoming refractory to remembering something than of forgetting. It serves to bring about brain states appropriate to the kind of response that is asked for in important instances of life: fear, thirst, sex, and hunger.

The presence of forgetting in our lives and in the lives of animals is so important that James McGaugh, a great worker in the area, once said that "forgetting is most salient part of memory" [1]. Thanks to the practice of that art, consciously or not, our collection of memories consists more of fragments and of extinguished or half-extinguished memories than of real and complete memories. But the judicious use of the art of forgetting plus that of imagination allowed some of the most fortunate among us, like Gabriel García Márquez, Borges, or Verdi, to produce masterpieces. By the way, García Márquez was already in the beginnings of what later was found to be Alzheimer's disease when he wrote his extraordinary autobiography mentioned above a few years ago.

© Springer International Publishing Switzerland 2015
I. Izquierdo, *The Art of Forgetting*, DOI 10.1007/978-3-319-06716-2_3

I hope I made it clear throughout the book that what most people call "forgetting" (meaning by that the difficulty or outright impossibility to retrieve memories *as if* they were lost) may in many cases be just the actual inhibition of retrieval, and not the erasure of memory traces. Maybe this fact is at the root of the fallacy attributed by some to psychoanalysis that "the brain cannot forget; all that gets in stays in." There is a considerable turnover of proteins and neurons in the brain; they are not made of indelible material. Each neuron changes its constitutive material many times a day. So much of the information that comes into the brain does not stay in: it is deleted and may or may not be replaced by others. Most of the information we acquired yesterday or last month or when we were 12 is gone forever. Some of it did not even make any stage of consolidation, and the rest of what was lost was effectively and definitively erased. The memories that were made and stayed in our brain may or may not be retrieved, depending on the operation of the processes of habituation, extinction, differentiation, or repression. Some of them may be falsified.

3.1   Final Comments

The major objectives of this book were to show that:

1. Losing memories can serve both the purpose of making space for new memories and to get rid of information that is no longer useful or otherwise unwanted.
2. Most memories are effectively lost without our awareness during daily life.
3. There are ways of getting rid of memories that are popularly called "forgetting" but that in reality involve just the inhibition of responding: extinction, habituation, and discrimination learning.. Eventually these processes may lead to the prolonged disuse of the synapses that store or retrieve those memories, which might on occasions lead to their atrophy and real forgetting.
4. In addition, I wanted to make the point that no doubt Norberto Bobbio was right to say that "we are what we remember," but we can add that "we also are what we chose to forget." Choosing to forget tells as much about us as what we remember. Like making memories, choosing to lose them involves a complex interaction between our will and our brain, which are usually in agreement, but not always.

Reference

1. Harlow H, McGaugh JL, Thompson RF. Psychology. San Francisco: Albion; 1971.

Index